D0960665

THE SCIENCE OF HUMAN PERFECTION

NATHANIEL COMFORT

The Science of Human Perfection

HOW GENES BECAME THE HEART OF AMERICAN MEDICINE

Yale UNIVERSITY PRESS

NEW HAVEN & LONDON

Published with assistance from the foundation established in memory of Philip Hamilton McMillan of the Class of 1894, Yale College.

Yale University Press books may be purchased in quantity for educational, business, or promotional use. For information, please e-mail sales.press@yale.edu (U.S. office) or sales@yaleup.co.uk (U.K. office).

Designed by James J. Johnson.
Set in Scala type by IDS Infotech Ltd., Chandigarh, India.
Printed in the United States of America.

Library of Congress Cataloging-in-Publication Data

Comfort, Nathaniel C.
The science of human perfection : how genes became the heart of American medicine / Nathaniel Comfort.
p. cm.
Includes bibliographical references and index.
ISBN 978-0-300-16991-1 (cloth : alk. paper)
I. Title.

[DNLM: 1. Genetics, Medical—history—United States. 2. Eugenics—history—United States. 3. Heredity—genetics—United States. 4. History, 20th Century—United States. 5. History, 21st Century—United States. QZ 11 AA1.]

616'.042—dc23

2012005081

A catalogue record for this book is available from the British Library.

This paper meets the requirements of ANSI/NISO Z39.48–1992 (Permanence of Paper).

10 9 8 7 6 5 4 3 2 1

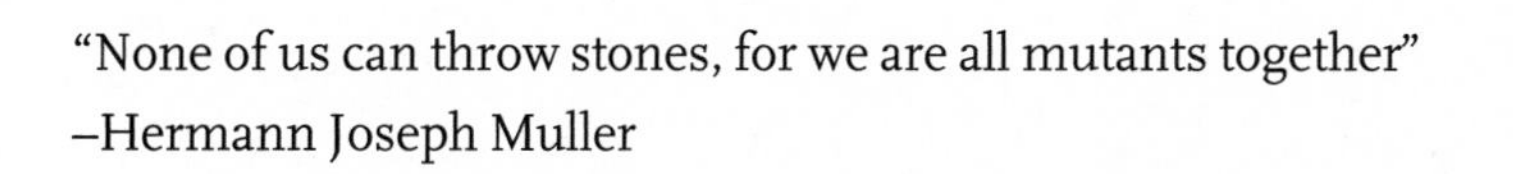

"None of us can throw stones, for we are all mutants together"
–Hermann Joseph Muller

CONTENTS

PREFACE

THIS IS A HISTORY OF PROMISES. Scientists, physicians, and reporters tell us that DNA science will enable us to live longer and to know when we will die. We will predict the diseases we will get and forestall them with drugs or lifestyle changes. Medicine will become "personalized": our doctor will know us literally outside and in, and so will be able to tailor medical care to our individual needs. We will know whether we are idiosyncratically sensitive or insensitive to a drug and whether that drug will be beneficial, inert, or toxic to us. Medicine will become a predictive, preventive science. We will simultaneously receive better care and have more control over and autonomy in our healthcare decisions. Ultimately, we will be able to engineer ourselves to eliminate disease, disability, and weakness, and we will guide our own evolution.

Although these promises have a glossy contemporary shine, they are recycled. Medicine has been going genetic—and genetics going medical—for more than a century. The claims and promises of genetic medicine predate DNA, the mapping of human genes to chromosomes, even genes themselves. It is not my task to evaluate whether these promises are realistic or can be kept. My goal is to uncover their origins, to illuminate their context, and to explain their meaning.

Medical genetics—or, more broadly, medical *heredity*—has always stemmed from two impulses: toward the relief of suffering and toward human improvement. The first is compassionate, medical, individual; to oppose it is misanthropic. The second is more complex. In the abstract, human improvement may be a laudatory goal, but hereditary human improvement denotes changing the nature of the population. It carries suggestions of human engineering, and of the objectification of the engineered. It is fashionable but disingenuous these days to assert that the relief of individual suffering is noble and population improvement insidious. Both are in fact both. Relief of suffering is humanitarian but raises issues of paternalism and autonomy. Deliberate efforts at human improvement may have the population's best interests at heart, but willy-nilly they involve social control.

The history of medical genetics is usually related as part of the larger history of human genetics. Over the years, a master narrative of human genetics has crystallized, in which the study of human heredity evolves from a focus on human improvement to one on relief of suffering. Medical genetics emerges out of this "medical turn" in human genetics, which occurred at the mid-twentieth century, in the years following the Second World War. This narrative allows one to acknowledge the distasteful origins of human genetics in eugenics—often defined in the words of the Progressive-era geneticist Charles Davenport as "the science of human improvement through better breeding"—while distancing oneself from it. The eugenic origins of human genetics are sequestered in the past, separated from modern medical human genetics by the historical firewall of World War II. Historians of eugenics have nuanced and deepened this narrative of a medical turn in human genetics, but they have largely retained its architecture. Years of scholarship have transformed the firewall into a membrane, multiply pierced with postwar eugenic practices and affiliations. Such work adds needed context, reflectiveness, and conscience to the often overamplified discussions of the promise of scientific medicine. Yet even with all this welcome texture, eugenics still tends to be construed as a "bad idea," even a pseudoscience, that has an insidious tendency to leak back into legitimate attempts to understand the hereditary basis of health and temperament. Eugenics is treated as exceptional, a scientific/social/political movement bound in time, an unfortunate phase in the history of human genetics. There remains in much of the scholarly discussion a sense that eugenics is a contaminant of good, honest biomedicine.

In contrast, in this book I treat human improvement and the relief of suffering as the two goals of all eugenics—and all medical genetics. Where previous accounts have documented the medical turn in human genetics, I trace, rather, a thin medical thread in studies of human heredity. It reaches back to the beginning of the twentieth century, gradually thickens, and comes to dominate scientific approaches to human heredity. Even at the beginning of the century, advocates of hereditary health made promises identical to those we hear today: genetics would make us healthier, longer-lived, smarter, happier—better. These goals were eugenic goals. I am interested less in the eugenics movement, which is historically situated in the Progressive era and interwar period, than in the eugenic impulse, which is timeless. It is the urge toward selection of the best offspring possible, toward the elimination of hereditary disease, and toward human engineering—"the self-direction of human evolution," as one Progressive-era poster put it. By focusing on the medical dimension of human heredity, we can see that these are constants in the history of biomedicine. Those constants find new expression in each era of our history. They have to be understood in their own context. This perspective dissolves the sense of inevitability and progression from our narrative of medical genetics and breaks our preoccupation with state control. Contemporary genetic medicine is not the fulfillment or completion of the science's promise but rather another stage in the evolution of a field that cannot exist outside of a social and cultural context.

I follow a small, evolving community of American scientists and doctors through the twentieth century, from the first stirrings of Mendelian genetics after 1900, through the building of a profession of human genetics around midcentury, up to the science-fiction achievements of recent biotechnology and the Human Genome Project. They are a diverse group of eugenicists, psychologists, geneticists, zoologists, physicians, statisticians, and public health workers. Many of these men and women will be unfamiliar even to scholars in the field; few are household names. I have traced their stories and contributions through their published work, both technical and popular; by reading their mail, memoranda, and other archival documents, as well as newspaper and magazine articles, biographies, and obituaries; and, for the more recent figures, by talking to them and those who knew them. Though their values and skills were diverse, they shared the belief that they could and should use their knowledge of heredity for the benefit of individuals, families, and society. They sought to integrate the science of heredity with the humanitarian aims of public health and medicine. A utopian spirit pervades their words and work. Heredity, they believed—and believe—is the foundation of human identity. Understanding it, manipulating it, controlling it can improve our lives.

As we follow them, medicine becomes genetic and genetics becomes medical. The narrative of the medical turn in human genetics implies that genetics colonized medicine; physicians seem almost passive receptacles for the genetic knowledge that sharpens and mechanizes their conception of disease. But I show that physicians actively imported genetics into their discipline—slowly at first, then with increasing vigor. As I tell this story, eugenics takes on a new role. It is no longer a sticky, noisome residue to be scrubbed off the skin of human genetics before it can go medical. Rather, eugenics is the lifeblood of medical genetics, the very reason genetics appealed to physicians. Science and medicine are equal partners in the formation of this new field. Medical genetics emerges as a true hybrid, not a graft. Another reading of this story, then, is as a case study in the history of biomedicine, with the tensions, negotiations, and alliances between the competing styles and interests of the scientist and the clinician. The hybridization of science and medicine, of course, is one of the defining characteristics of twentieth-century healthcare.

This reading of American medical genetics has two subversive effects, two sides of the same coin. First, by tracing the medical thread through early human genetics, I reveal that period to be less confused and malicious than it has often seemed. The first years of human genetics was not merely "sloppy science" and racist dogma. Much of the medical side of early human genetics was aimed at the same basic goals as genetic medicine today—and even at many of the same targets, of health, personality, and intellect. Obversely, contemporary genetic medicine emerges as being less benign than the public relations campaigns would have us believe. The desire to relieve suffering and to improve ourselves genetically is noble but freighted with social and ethical consequences. The promises of genetic medicine *are* the promises of eugenics.

The genetic approach toward health has enormous power to reduce suffering and improve our lives. But it is more than an altruistic endeavor: it is also a fascinating set of puzzles, a powerful political tool, and big business. Historicizing the promises of genetic medicine allows us to critically explore the interplay among the economic, political, intellectual, and humanitarian impulses driving genetic medicine. As biomedicine and healthcare become increasingly important in daily life, a healthy skepticism becomes literally vital. It can help us benefit from the powerful new knowledge biomedicine daily produces. As patients and consumers, we must use that knowledge intelligently—lest other interests trump our own.

ACKNOWLEDGMENTS

TWO MEN NO LONGER LIVING REQUIRE pride of place in my thanks. I first realized I needed to write about medical genetics when, visiting the Johns Hopkins medical school in 2001, I got an appointment with Barton Childs. After forty-five minutes with him, I knew what my next book would be about. He met with me several times after that and answered my many ignorant questions, usually patiently. My view of genetic medicine has evolved quite a lot since then, and not always in directions he would agree with, but I feel I learned about the Garrodian approach at the feet of the master. Soon after meeting Barton, I was introduced to Victor McKusick. He, too, took me under his wing, met with me often, and helped me understand his approach. He was a true physician-scholar, generous and gracious, and I am grateful for his and Anne McKusick's kindness. Joining the Hopkins medical school faculty gave me access to them and to many other scientists who suffered, apparently gladly, questions that must often have seemed out of left field. In particular, David Valle, an ardent Garrodian (and Childsian) and Aravinda Chakravarti, who considers himself both a Garrodian and a Galtonian, gave me much advice and philosophical conversation, many editorial comments, and pretty decent coffee. I am indebted to these scientists and physicians and have been most fortunate to work in a place where I am surrounded by my historical actors.

The many scientists and physicians I interviewed have been extraordinarily kind and hospitable. My interviews are pretty grueling affairs, eight hours or more, stretched over two to three days. These men and women were at least tolerant of and often enthusiastic about the process. They often invited me to their homes, feeding me, even putting me up at night. Their influence goes far beyond the quotations found in these pages. Among them (besides those mentioned above) are Margaret Abbott, Peter Harper, Haig Kazazian, Alfred Knudson, Barbara Migeon, Arno Motulsky, Walter Nance, Peter Nowell, Frank Ruddle, Charles Scriver, and Oliver Smithies. Early in this project, I joined Edward McCabe and Marcia Meldrum of UCLA in creating the Oral History of Human Genetics Project (ohhgp .pendari.com), which has shaped this project in many ways. The project received funding support from both the National Institutes of Health (R01HG003206–01) and the National Science Foundation (0551068).

Many friends and colleagues read all or part of the manuscript and/or engaged with me. First among them is Alex Stern. It has been great fun and solace to have complementary books progressing at exactly the same pace, right down to the final deadline. Robert Resta has become a valued colleague and friend. Angela Creager, Kathy Cooke, Michael Gordin, Diane Paul, Susan Lindee, and Audra Wolfe have all given me valuable comments and support. Seminar audiences at Hopkins, Princeton, Yale, Arizona State, the University of Pennsylvania, the University of Michigan, and the Oxford College of Emory University peppered me with tough questions that helped mold naïve ideas into something a little deeper. The audience in my own program's colloquium, including all our faculty and students, was particularly helpful at a crucial stage. Dan Todes gave me both a critical reading and many helpful bourbons. Equally important are my "educated lay" readers, Richard Comfort, Keith Hale, and especially Michiko Kobayashi, who is a good reader, a good listener, and a good thinker, and who has a knack for a well-timed shoulder rub. Thanks to my chairman, Randy Packard, for deferring the exigencies of academic life as long as possible.

An undergraduate student, Peter Lippman, was a faithful and diligent research assistant on the middle part of the book. Archivists at the various institutions I worked at were extremely helpful and generous, particularly Clare Clark, Monica Garnett Coffey, Charles Greifenstein, Andy Harrison, Phoebe Latocha, and William K. Wallach. Research at the Bentley Historical Library at the University of Michigan was aided by a Bentley Travel Grant.

Gavan Tredoux answered several technical points and helped me find images. Jean Thomson Black at Yale University Press has been a most supportive and efficient editor, and Dan Heaton fixed my danglers and improved the manuscript in many ways. Three reviewers, two anonymous, were extremely helpful in sharpening the book.

In different ways my parents, Richard Comfort and Louise Comfort, provide much-appreciated support; they are good readers and thoughtful critics. Continuing the tradition, my children, Charles and Gwendolyn, listen to my stories, understand when I have to work instead of watch movies, and above all make me laugh.

1

The Galton-Garrod Society

IN ABOUT 1950, FOUR RESEARCHERS AT Johns Hopkins University did a very ordinary, very significant thing: they started a journal club. For decades, journal clubs have been a staple of laboratory life; every basic scientist I know attends at least one. They are informal groups, often with no criteria for membership beyond interest. Science being the focused enterprise that it is, the clubs usually draw from one or two laboratories. Occasionally a colleague from a neighboring department is invited to join. Typically, they meet weekly or biweekly, reading a recent paper, then convening to discuss it. On a rotating schedule, one member leads the group in critically dissecting the paper, figure by figure. Forming a journal club is thus a commonplace; but this one was special.

First, it was interdisciplinary. Victor McKusick was from the Department of Medicine, Barton Childs from Pediatrics, both in the School of Medicine. Abraham Lilienfeld was from Epidemiology, in the School of Public Health and Hygiene. H. Bentley Glass was from the Department of Biology, in the School of Arts and Sciences, across town on the undergraduate campus. Four departments and three schools of the university. Such broad cross-pollination is almost unheard of today.

Second, the subject of the club was medical genetics, an exotic and sensitive theme. The diverse backgrounds of the group's members reflected the range of interests and expertise that have gone into medical genetics. It's a specialty that crashed the national borders distinguishing medicine, science, and public health, and crossed many state lines within each of those nations. Also, it still slightly upset the border patrols. Because of its association with Progressive-era eugenics, medical genetics still carried a whiff of suspicion among physicians, particularly at top-tier schools like Hopkins. It took some courage to form the group.

Third, although the journal club wasn't the start of medical genetics at Hopkins, it was the moment of nucleation, the first stirring of anything that looked like *organized* medical genetics at Hopkins. In the following years, the Johns Hopkins medical school became a hub for the field, which a few years hence expanded and professionalized rapidly, evolving into contemporary genetic medicine, including the Human Genome Project, gene therapy, biotechnology, and large sectors of the pharmaceutical industry. The club thus looks like an origin moment, however symbolic, for a discipline that has an enormous impact on our lives today.

But fourth, looking backward, the club's formation can be interpreted as the integration of historical and intellectual traditions stretching back to the nineteenth century. It is therefore better read as a node, a point of convergence, than as an origin. As such, it gives insight into that which came before, as well as that which followed. We begin, then, at the temporal center of our larger story.

The Hopkins researchers named the club historically, with an eponym that can stand for the central theme of this book. They called their group the Galton-Garrod Society. "Galton" referred to Francis Galton, a cousin of Charles Darwin, a polymath, a pioneer in statistics, and well known as the person who coined the term *eugenics*. "Garrod" was Archibald Garrod, a younger contemporary of Galton's and like Galton a Londoner. Garrod was a pediatrician interested in biochemistry who—in collaboration with the English naturalist William Bateson—described the first Mendelian trait in human beings. In naming their reading group after Galton and Garrod, the Hopkins researchers identified the two men as the twin founders of medical genetics. This seems to me astute and revealing, both in ways the Hopkins researchers intended and in ways they certainly did not.[1]

I want to use Galton and Garrod as totems of the paired, opposing forces whose interactions, I suggest, have shaped the contours of medical genetics as it developed over the twentieth century. They are the two sides in a dialectic:

the tension between them is like a spring, storing the driving energy that propels the field. As the chapters unfold, we will see how the Galtonian and the Garrodian tumble down the decades, evolving out of Progressive-era eugenic human genetics and eventually dissolving into each other, producing modern genetic medicine. By way of introduction, then, let us define *Galtonian* and *Garrodian* through brief looks at the lives of their namesakes. We will also need a detour, to dip into the life and work of Bateson, Garrod's crucial collaborator, and into the origins of the Mendelian genetics without which neither Garrodian nor Galtonian thought is comprehensible.

* * *

Francis Galton has an oddly polarized legacy. Some biographers revere him as an eclectic genius and a pioneer of population biology. Yet others have treated him as something of a buffoon—on the cover of Martin Brookes's recent biography, for example, Galton reclines in a bathtub, absurd and undignified. Both poles capture some of the truth.[2]

Born in 1822 near Birmingham, England, he was the last of seven children. His mother, née Violetta Darwin, was a sister of Robert Darwin, Charles Darwin's father. The young Darwin had been a happy and unremarkable child, a mediocre student at best. His cousin Francis Galton, on the other hand, was a prodigy. He might have been merely bright, had not a zealous and bored invalid sister coached his native gifts to remarkable heights. He was a poster boy for the interaction of nature and nurture.

As Robert Darwin had done to his sons Erasmus and Charles, Galton's father pushed him into medicine—and like his cousins, Galton ultimately rejected it. He studied medicine at Birmingham General Hospital and then at King's College in London, before enrolling at Cambridge in mathematics. He studied obsessively. When his stamina faded, he unlimbered his Gumption Reviver, a large funnel that dripped water on his head at an adjustable rate to keep him awake. Under the strain of his intense study, he experienced repeated breakdowns, partly as a result of his perfectionism and anxiety to please his father. Galton was crushed when he failed to earn a first, taking an ordinary degree in 1844.[3]

The death of his overbearing father later that year was a double liberation: it removed the pressure on Galton to become a doctor and provided him with a considerable inheritance. Travel writing (his *The Art of Travel* went through eight editions), exploration, and such related topics as meteorology and navigation were the primary themes of the first half of Galton's career. Increasingly, however, his interests turned toward heredity and the measurement of human traits.

By 1857, the year Garrod was born, Galton was a fixture of British scientific society. Over the subsequent five decades, he published more than four hundred books, articles, and pamphlets. As we parse intellectual disciplines today, he explored meteorology, psychology, anthropology, sociology, criminology, evolution, eugenics, and statistics. The line distinguishing the polymath from the dilettante is fat and fuzzy, and Galton straddled it. In 1883 George Romanes, the well-known disciple of Darwin, wrote, "Mr. Francis Galton has no competitor in regard to the variety and versatility of his researches"; in fact, his interests ranged so widely that one tended "to regard them as disconnected pieces of work, which from time to time were thrown off like sparks from the flame of an active mind." But with the perspective of a century and a half's distance, we can in fact see some thematic clusters in Galton's efforts.[4]

Perhaps more than anything else, he was an inventor and a tinkerer. All his life he invented gadgets, tricks, shortcuts, and novelties. Among them: underwater reading glasses for divers, a printing teletype machine, a bicycle speedometer, a hand heliostat for signaling over long distances, a rotary steam engine, a device for measuring vapor tension, composite photography, several "pocket registrators" for conducting surreptitious anthropological studies, numerous scales, metrics, and devices for gauging human traits and qualities, and the quincunx, a sort of pinball machine for demonstrating the normal, or Gaussian, distribution. Many of his inventions were crude. Some were obvious. But several were clever and a few were brilliant. He invented the modern weather map, with its now-familiar nested lines indicating regions of equal temperature or barometric pressure. He developed fingerprinting as a modern tool of criminology. And he developed numerous tools, both physical and statistical, that became mainstays of human population genetics. Galton could be simplistic, and he often got mired in detail, but he was no fool.

His experimental work was similarly eccentric. Among his published articles one may find "Statistics of Mental Imagery"; "Head Growth in Students

1.1 Francis Galton, aged forty-two, at the height of his powers. From Pearson, *Life, Letters, and Labours of Francis Galton*, vol. 2, plate IX, between 67 and 68

at the University of Cambridge"; "Sun Signals to Mars"; "Arithmetic by Smell"; and "Three Generations of Lunatic Cats." Intellectually, Galton was nearly the opposite of his cousin Darwin. Where Darwin was a meticulous collector whose written corpus moved carefully, book by book, toward a sweeping synthesis of natural history, Galton went around day by day making curious observations and solving countless problems, large and small.

Galton was a great quantifier. Many of his inventions were measuring devices. "Whenever you can, count" was his motto. He was a better measurer than calculator—despite his Cambridge degree in mathematics, he needed help with any sophisticated analysis and he often made mistakes in his own work. He was more precise than accurate. He had a knack for taking a sensitive, even poetic subject and desiccating it with numbers—the sensitivity of the nape of a woman's neck, for example, measured and quantified and parsed into percentiles. He had, however, an unusual talent for spotting trends in linear data.[5]

What Galton liked to measure most were human traits and qualities. He explored the temperaments and turns of mind that characterize people of different degrees of social success. He systematized the whorls and waves of human fingerprint patterns as a means for the authorities to keep track of social deviants. In 1884 he set up an Anthropometric Laboratory at the International Health Exhibition, where he measured everything from height and weight to reaction time and color sense on thousands of volunteers. But he was at least as interested in corralling with quantification such unruly traits as temperament, imagination, and spirituality. His most controversial study was a statistical examination into the efficacy of prayer, a behavior that, as a committed evolutionist, he found difficult to comprehend. He then compiled these masses of data, often devising novel methods of visualizing them through tables, charts, and graphs. From them—largely intuitively, it seems—he sought to extract the truths of human nature.[6]

For Galton, truth was smooth and lawlike, once one stripped away all confounding individual variation. The signature of his thought is the normal or Gaussian distribution—the familiar bell curve. In biology, many complex traits will display a normal distribution: most individuals fall close to a mean value, with sharply fewer individuals at either extreme. Height in humans is a classic example of a normally distributed trait; eye color, in contrast, is discretely distributed over several distinct colors. Galton was passionately interested in variation within a population, but he was singularly uninterested in individuals.

He had enormous faith in the power of statistics to express these essential, ideal truths. Statistics—so named because originally it was used to compute state actuarial tables—was a new science, growing rapidly in its explanatory power. Galton greatly expanded the vocabulary of statistics in biology, and it is on these contributions that his reputation as a founding figure of population genetics primarily rests. The core data set for his 1889 *Natural Inheritance* was a survey, the Record of Family Faculties. The returned questionnaires, which he somewhat confusingly also called "records," listed data on all available family qualities, from talents to physical characteristics to diseases and causes of death. Galton awarded cash prizes for each record received, depending on the quality of the data. He received about 150. The collection of Galtonian records was among the most important techniques of human genetics until well into the second half of the twentieth century.[7]

Analyzing his Record of Family Faculties data, Galton found that for a given trait offspring tended to be less extreme than their parents; they "regressed" toward the average value, or mean. This seemed to him a profound truth of biology, and it guided much of his later thinking about heredity. As a biological principle, regression is an artifact (tall children, for example, tend to fall between their parents, not between their parents and the mean), but regression analysis generalized into one of the most fundamental tools of statistical analysis. It allows one to find the best straight line through a cloud of points on a graph. Out of the same data set, Galton developed an expression for finding the extent to which one variable depends on another. Galton's student and biographer Karl Pearson developed this into the statistical technique of correlation.[8]

Galton's commitment to the principle of regression to the mean led him to a conundrum. If offspring always tended to fall between their parents for a given trait, how could evolution ever occur? Answering this question was essential not only logically but socially as well, for Galton was passionate about using knowledge of heredity to improve society. Did regression imply that society was doomed to homogeneous mediocrity? The only way he could see out of this dilemma was through "sports"—what we would call large-scale mutations—coupled with selection. An occasional individual with an extreme value of a trait would pull the mean upward. Thus Galton came to favor discontinuous or saltational evolution, in contrast to Darwinian gradual change.

His rejection of Darwinian gradualism, resulting from disagreement over heredity, was painful. In 1871 Galton devised a set of transfusion experiments with rabbits that was designed to support pangenesis, Darwin's ill-considered theory of heredity, put forward in 1868. Pangenesis asserted that the body's tissues "throw off" particles that circulate in the body and gather in the reproductive organs to form the hereditary material. But when Galton transfused blood from the ear of one type of rabbit to another, Darwin's hypothetical hereditary particles failed to cross over; the animals' progeny looked like their parents, not the other breed. Darwin cried "foul": he claimed that he had never said the particles circulated in the blood. But in Galton's mind Darwin's theory was disproved. In place of pangenesis, Galton suggested that the hereditary material passes down through the generations untouched by the organism's experience. He called this the stirp—from the Latin for "root." Today we use the term *germ line,* in distinction to the *soma,* or body. The distinction between "nature" and "nurture"—Galton may have borrowed the pairing from Shakespeare's *The Tempest*—remained a central theme in Galton's work and, of course, in all subsequent discussion of the social meaning of genetics.[9]

One of his favorite methods for distinguishing nature from nurture has proven robust. In 1875 he wrote in *Fraser's* magazine that comparing traits in pairs of fraternal and identical twins offered "a means of distinguishing between the effects of tendencies received at birth, and of those that were imposed by the circumstances of their after lives; in other words, between the effects of nature and nurture." The study of twins helped persuade Galton of the dominance of heredity over environment. "There is no escape," he wrote, "from the conclusion that nature prevails enormously over nurture" if one controls, even loosely, for social status and culture. Elsewhere, he wrote, "I look upon race as far more important than nurture." Such hereditarianism is a hallmark of Galtonian thought and one of his most lasting contributions to the study of heredity.[10]

The beauty of this hereditarian outlook, for Galton, was that it rendered human nature malleable. "It would seem as though physical structure of future generations were almost as plastic as clay, under the control of the breeder's will," he wrote. So too were mental qualities "equally under control." This plasticity of constitution meant that populations were more easily changed than individuals. "The human race has a large control over its future forms of activity,—far more than any individual has over its own." Humankind, he wrote, "can modify its own nature." We must not be misled by the notion of "individuality," he cautioned. "Our personalities are not so independent as our self-consciousness leads us to believe." Galton preferred to think of individual humans as part of a larger cooperative whole, in solidarity with one another and each willing to sacrifice for the common good.[11]

So Galton was a hereditarian idealist, with a deep belief in the unity of the organic world. "All life is single in its essence," he wrote. "All men and all other living animals are active workers and sharers in a vastly more extended system of cosmic action than any of ourselves, much less of them, can possibly comprehend." This unity led him to fantasize about perfecting human society, and therefore humankind. "Let us give reins to our fancy," he wrote in 1865, "and imagine a Utopia." And again, thirty-six years later: "It is pleasant to contrive Utopias, and I have indulged in many, of which a great society is one." Throughout his writings, he seems unaware that utopias always have a dark side; even in imagined societies there is always a cost, and it is usually liberty. Of course, thinkers at least back to Plato had fancied hereditarian utopias, but Galton's was a veiled scientific treatise, a germ of a biosocial movement.[12]

Out of these qualities, both constitutional and learned, came Galton's hereditarian, population-based strategy for human improvement. Both Galton and Darwin staunchly believed that humans and other organisms were subject to the same laws of heredity and evolution. But whereas Cousin Darwin had said that nature acts like a slow pigeon fancier, excruciatingly lazy and lacking foresight, Galton wondered why we who invented artificial selection cast ourselves to the inefficient whims of nature. He first floated the idea for engineering society in 1865, in an article for the general-interest magazine *Macmillan's*. He fantasized about an entire race of idealized upper-class English intellectuals: brilliant, cultured, genteel, restrained, and conscientious. To get there, he invoked an image of animal husbandry that became a standard trope of American eugenicists in the twentieth century: "If a twentieth part of the cost and pains were spent in measures for the improvement of the human race that is spent on the improvement of the breed of horses and cattle, what a galaxy of genius might we not create!" Initially, he called the idea "viriculture," but by 1883 he had settled on the less agricultural *eugenics,* meaning "well born."[13]

Despite metaphors of animal husbandry, Galton was careful to stipulate that eugenics need not be coercive. His idealism gave him great faith in education and reason; he believed that people, acting upon "existing conditions of law and sentiment," would voluntarily regulate their mating practices for the good of the population. His plan was "perfectly in accordance with the moral sense of the present time," he wrote in 1873. Indeed, in 1904 he explicitly denounced the coercive breeding that some Progressive eugenicists were already beginning to advocate. "We can't mate men and women as we please, like cocks and hens, but we could I think gradually evolve some plan by which there would be a steady though slow amelioration of the human breed; the aim being to increase the contributions of the more valuable classes of the population and to diminish the converse." Fulfilling this aim required further research into the traits of interest. He continued, "We now want better criteria than we have of which is which." In short, Galton would improve the hereditary stock indirectly, by acting on the environment: by means of educational pamphlets and books and, he hoped, incentives such as tax breaks for the wealthy and successful.[14]

One can move the bell curve to the right either by increasing the numbers at the high end (positive eugenics) or by decreasing the numbers at the low end (negative eugenics). Thus positive and negative here are statements of mathematics, not value—positive eugenics is not "better" than negative. Over the years, each has been in vogue (at the beginning of the twenty-first century, most people consider elimination of hereditary disease more acceptable than genetic enhancement, though this may be changing). Galton was more interested in positive than negative eugenics. "Increasing the productivity of the best stock," he wrote, was "far more important than that of repressing the productivity of the worst."[15]

One important aspect of the "worst" was disease. Galton devoted a chapter of *Natural Inheritance* to the subject. He spent the majority of it not on hereditary disease—of which several were known—but on an illness known to be contagious: consumption, which corresponds roughly to tuberculosis. He focused, therefore, on a disease important to Londoners, rather than one that suited his method. He concluded that quantitative evidence accords with common sense: that consumption is largely acquired, except where physical malformations predispose one to the disease; that a mother's tendency to transmit a consumptive taint to her children is mostly due to infection; and that "consumptivity," the constitutional predisposition to consumption, is both real and inherited. Five years later, some of the most distinguished physicians in America arrived at similar conclusions. Galton was among the first to try seriously to tease out the hereditary component of complex disease.[16]

In most ways, though, Galton was a world apart from the medical perspective. His emphasis on the population drained him of compassion for individuals. Statistically, individual variation was noise that had to be filtered in order to discover the pure underlying truth of general trends. And philosophically, reason, not sentiment or virtue, was for him the proper infrastructure of Utopia. A reasoned solution to a social problem would be free from favoritism or sentiment and would therefore do the greatest good for the greatest number. In his mind, eugenics was a means of achieving a "great society." Interviewed by the *Jewish Chronicle* in 1910, he claimed that persecution could strengthen a race by selecting for the hardiest, most resilient individuals. When asked whether this was not rather immoral, he denied the charge, calling his position "unmoral" instead. It is common for a scientist to claim that scientific objectivity excuses him from, or indeed precludes, considering the moral implications of his work. And yet it is not hard to believe that to an emotionally distant man such as Galton, eugenics would seem an expression of kindness. "It is the aim of eugenics," he continued, "to supply many means by which the effects of these drastic and not always successful aids to race culture [that is, persecution] may be produced in a more scientific and kindly way." Eugenics, in other words, would have all the biological benefits of persecution, without the unpleasant moral odor.[17]

Thus Galton's talent, hereditary and otherwise. His principal contributions to human genetics were to develop and popularize several methods that became important to the field, his articulation of the distinction between nature and nurture, and, overarchingly, his population-based approach to human heredity. The methods he pioneered—questionnaires, twin studies, and statistics—were mainstays of human genetics for more than a century, and in modified form they still compose part of the backbone of human-genetic research. His mind percolated with thousands of ideas, some whimsically eccentric, others penetrating, but all austere and oddly cool; he was a man alone with his toys. Eugenics embodied all that he stood for.

Galton's inheritance, investments, and royalties ensured that he died a wealthy man. Yet in one respect he was poor: the father of eugenics left no heirs. He died in 1911, at the age of eighty-nine. By this time, eugenics had become popular, thanks in large measure to the rediscovery of Gregor Mendel's law of heredity in 1900. Where Galton sought to explain normally distributed, continuously varying traits like height, Mendelism explained discrete characters like eye color. Indeed, an impassioned feud broke out between the Mendelians and Galton's acolytes, the biometricians. Nevertheless, Mendelism gave credence to Galton's unshakable belief in the primacy of nature over nurture. The popularity of eugenics after 1900 was such that Galton could establish a fellowship in eugenics, a National Eugenics Laboratory, a scientific journal, and, with a bequest, an endowed professorship of eugenics, at University College, London. And ironically, at midcentury, the Galton Laboratory became the cradle of Garrodian medical genetics.

* * *

In 1897 a London woman gave birth to what appeared a perfectly normal baby. But soon after the delivery, the mother made a horrific finding. Changing a diaper, she discovered it was brownish-black. Doubtless envisioning morbidity's residue leaking out of her infant's urinary tract, she grabbed baby and diaper and headed for the Hospital for Sick Children, in Great Ormond Street. Dr. Garrod was on call that day.

Archibald Edward Garrod was forty-one years old, a serious, cool man, with a bushy, drooping mustache and a piercing gaze under a heavy brow. The son of a biochemically minded physician, Garrod had grown up amid his father's beakers, flasks, and retorts; he later recalled the crystallizing dishes on the mantel in which the elder Garrod had collected and purified the uric acid for his classic study of gout. Though undistinguished in his early education, Garrod had entered Christ College, Oxford. He began to thrive intellectually. In 1880 he took a first in the Natural Sciences Tripos.[18]

Like his father, the younger Dr. Garrod was interested in chemistry and its role in medicine. He had studied in Vienna, and in the nineties he collaborated with Frederick Gowland Hopkins, the renowned London physician and biochemist who pioneered the study of dietary deficiency and the "accessory factors"—later, vitamins—that prevented them. Dr. Garrod's main appointment was at St. Bartholomew's, London's premier teaching and research hospital, where he was a demonstrator in chemical pathology, but he also had positions at several other area hospitals, including Great Ormond Street.

Since working with Hopkins, Garrod had been fascinated with urine, especially when it turned different colors. He had published papers with titles such as "On Haematoporphyrin as a Urinary Pigment in Disease," "The Spectroscopic Examination of Urine," and "A Specimen of Urine Rendered Green by Indigo." Uroscopy has a long and unsavory history. Once a celebrated diagnostic tool, by the sixteenth century it had become an object of derision among learned physicians. A scathing seventeenth-century pamphlet mixed no water with its vinegar: "The Pisse-Prophet; or, Certaine Pisse Pot Lectures" condemned the specialty's fraudulent practices. But Garrod was no Pisse Prophet. Where the urinary quacks had swirled a flask and made a diagnosis, he applied chemical analysis and recent techniques such as spectroscopy to gain clues to the body's internal mysteries.[19]

Dr. Garrod took great interest in the patient with the black diaper. He took a family history, and as the baby grew he kept in touch with the family. He sought out other patients with the condition. There weren't many; it occurs in one of four million people. He managed to scrape together forty cases, gathered from colleagues and published cases in the medical journals. He also read up on the history of the condition. Digging back through the literature, he found that black-urine disease was first described in 1822. In 1859 Carl Boedecker named the black pigment "alkapton." In 1886 Robert Kirk wrote that it seemed to run in families. Five years later, two Germans, Michael Wolkow and E. Baumann, described the biochemistry in a major and highly technical article. The pigment—alkapton, or homogentisic acid—comes from tyrosine, one of two amino acids in our diet that contain a benzene ring, a six-carbon chain that bites its own tail. Homogentisic acid also contains the benzene ring. Doctors believed that human cells could not make homogentisic acid. According to the experts, alkaptonuria must result from an invasion of the gut by microbes that could convert tyrosine to homogentisic acid. Germ theories were high fashion. Since the early 1880s, every year had brought the discovery of new disease germs. The germ theory of alkaptonuria was a natural guess, based on sound data interpreted in light of the best contemporary science.[20]

1.2 Archibald Garrod, the reluctant geneticist. Courtesy of The Royal Society

In 1900 the mother of Garrod's alkaptonuria patient became pregnant again. He followed her progress avidly. The baby was born on March 1, 1901. Garrod had the nurses check every diaper. Staining was first noticed fifty-two hours after birth—too soon to be the result of infection, Garrod judged. He concluded that alkaptonuria must be innate. Garrod began writing up a clinical report, which he read to the Royal Medical and Chirurgical Society in November and followed up with a brief article, "About Alkaptonuria," in the *Lancet,* one of the two major English medical journals. He called alkaptonuria a "freak of metabolism."[21]

One feature of the condition particularly caught his notice: it had a "special liability" to occur in the children of first cousins. More than one-quarter of his forty alkaptonurics were the product of so-called consanguineous marriage. In three of four families with alkaptonuric children that Garrod had data on, the parents were related. Garrod did not need fancy statistics to recognize that because "the children of first cousins form so small a section of the community, and the number of alkaptonuric persons is so very small," such a strong correlation could "hardly be ascribed to chance."[22]

Garrod's paper caught the eye of the London naturalist William Bateson. Born in 1861, in Whitby on the English coast, Bateson was artistic, moody, and drawn to nature and language. He once came up with a theory of heredity by staring at sand ripples on the beach. He has been called conservative, socially and scientifically, but he was, rather, a purist. He was a scientific aesthete, passionate and intellectually ruthless. The great theoretical biologist J. B. S. Haldane wrote that as a scientist, Bateson "never attempted to conceal his contempt for second-rate work or second-rate thought." As a schoolboy, however, his own work was second-rate—but this was for lack of passion, not ability. At Rugby, his marks strained toward but rarely attained mediocrity. His father, conveniently, was a master of classics at St. John's College, Cambridge. Bateson matriculated there in 1879, performed poorly in nearly every subject, and then surprised everyone when he took a first in the Natural Sciences Tripos in 1882. He took another first in part two, in zoology, the following year. He had found his métier in evolution. Like Garrod, Bateson studied abroad, but rather than east to Europe he went west, to the Chesapeake Bay of Maryland and Virginia, where he spent two summers working under the great marine biologist William Keith Brooks of Johns Hopkins University, mentor to some of the biggest names in American biology, including, a few years later, the geneticist Thomas Hunt Morgan.[23]

At Cambridge, Bateson befriended W. F. R. (Frank) Weldon. Studying together under the great morphologist Francis Balfour, Bateson and Weldon became best friends. The two men read both Darwin and Galton's *Natural Inheritance* differently, however. Bateson was a staunch Darwinian, but he favored Galtonian saltations as an evolutionary mechanism. Weldon was devoted to Galtonian biometry, but he preferred Darwinian gradualism. In 1894 Bateson published *Materials for the Study of Variation,* his masterwork on discontinuous evolution. To Galton's chagrin, Weldon reviewed it savagely. Bateson took it hard. They began a feud, both scientific and personal, that lasted until Weldon's death in 1906.[24]

In 1900 Bateson got his most important ammunition against Weldon: the hereditary theory of Gregor Mendel. In 1866, in Brno, in what is now the Czech Republic, the monk Mendel had published a paper describing what he called a principle of heredity. No one in the English or French natural history communities seems to have read the paper; it had no impact on the raging debates over heredity and variation in evolution. It languished, scarcely cited except in catalogues, for thirty-four years, until two Continental plant breeders, the German Carl Correns and the Netherlander Hugo de Vries, rediscovered the principles and the paper.[25]

Immediately Bateson saw Mendel's value to his cause. The monk's segregating particles seemed perfect for discontinuous evolution. If heredity were particulate, he believed (incorrectly), evolution could not be continuous. He immediately became the most ardent Mendelian of his day. In May 1900 he wrote that Mendel's principles "will certainly play a conspicuous part in all future discussions of evolutionary problems. It is not a little remarkable that Mendel's work should have escaped notice, and been so long forgotten." By the time he encountered Garrod in late 1901, Bateson was already writing a Mendelian manifesto. He translated Mendel's 1866 paper from German into English. Then he wrote a lengthy gloss on its implications. Finally, he composed a scathing attack on Weldon, in the course of which Bateson laid out the principles of the science of genetics. The result, *Mendel's Principles of Heredity: A Defence,* is a strange, sprawling book, by turns brilliantly incisive

1.3 William Bateson, Mendel's bulldog. Courtesy of the John Innes Centre

and maddeningly petty. It is part historical document, part tutorial, part treatise, and part assault. It is at once dirty-laundry personal and test-tube elegant.[26]

Bateson's greatest legacy is his coinage; from him comes much of the language of genetics. He gave the name *allelomorphs*—later shortened to *alleles*—to the two alternative forms of a Mendelian element. He defined an individual with two copies of the same allelomorph (one from each parent) as the *homozygote;* one with two different alleles as the *heterozygote.* Later, in 1905, applying unsuccessfully for an endowed chair in the field he had helped to invent, he named it, coining the word *genetics.*[27]

Bateson, said J. B. S. Haldane, was greater than any of his ideas. His theories were like iron knives—incisive initially, they dulled with use. His intellectual style was formal, abstract, aesthetic. For Bateson, the Mendelian elements—what we call genes—were not physical particles. Heredity behaves, he believed, only *as if* there were particles. He stuck to this view doggedly, until nearly the end of his life. He described a process of "coupling" and "repulsion" between elements, similar to what geneticists call "linkage," the association of two traits through the generations, like red hair and green eyes. Most important, he distilled Mendel's principles to their simplest form. Mendel had said there were two types of elements for each trait: dominant and recessive. In Bateson's version of Mendelism, there was only one type of determiner: the dominant. The recessive allelomorph, he said, was merely the absence of a determiner. This became known as his presence-and-absence theory. This ruthless distillation of heredity's mechanism to abstract principles gave Bateson's version of Mendelism enormous explanatory power—until mounting evidence made it untenable.[28]

* * *

No one in the English-speaking world, then, was more qualified to solve a hereditary problem than Bateson. When he read about Garrod's alkaptonuria patients, he recognized immediately that the condition followed a classic Mendelian recessive pattern. The two men struck up a correspondence. As Garrod peppered Bateson with questions, Bateson replied gamely and patiently, explaining the Mendelian principles and the patterns of recessive allelomorphs. The exchange lasted at least until June 1902. Thanks to Bateson, Garrod came to grasp the Mendelian ratios, the abstractions so foreign to a medical man of that day. Bateson in return was thrilled to prove that Mendelian heredity occurred in humans; this removed a weighty potential objection to his argument for discontinuous evolution, and supported beautifully his presence-and-absence theory.[29]

On the basis of Bateson's tutelage, Garrod wrote a third article, "The Incidence of Alkaptonuria: A Study in Chemical Individuality," which was published in the *Lancet* in 1902. It is often cited as the founding paper of biochemical genetics. But like Mendel's paper thirty-six years before, it first languished, almost unread, for decades. We may forgive physicians reading in 1902 for not recognizing Garrod's paper as a deathless classic. Although clearly written, it is often elliptical, Garrod taking for granted that his audience shared his wide knowledge of chemistry. The reasoning is dense, and crucial ideas can be missed if one's attention wanders. Read carefully, it is indeed a medical masterpiece. But squeezed among the other articles of the *Lancet,* it must have seemed yet another obscure article by yet another erudite physician.[30]

It begins as though the reader is breaking in on a medical cocktail party conversation: "All the more recent work on alkaptonuria has tended to show that the constant feature of that condition is the excretion of homogentisic acid." Alkaptonuria is so rare that few of Garrod's readers had ever seen a case of it. And what was homogentisic acid? Garrod continues by describing chemical experiments converting alkapton acid into its ethyl ester, determining the melting point of crystalline homogentisic acid, and other arcana. He establishes that homogentisic acid derives from the amino acid tyrosine. To appreciate the weight of his facts and reasoning, the reader has to know that the benzene ring is a hexagonal chain of carbon atoms, that it is toxic to humans, and that both tyrosine and homogentisic acid contain it. He has to know that snapping that ring is an essential step in breaking down, or metabolizing, protein in our diet. "Why alkaptonuric individuals pass the benzene ring unbroken and where the peculiar chemical change from tyrosin[e] to homogentisic acid occurs remain unsolved problems," Garrod writes. Such problems, relating to a rare and nonmorbid condition, were not of burning concern to many physicians.[31]

Further, Garrod was not interested in alkaptonuria as a disease. "There are good reasons for thinking that alkaptonuria is not the manifestation of a disease but is rather of the nature of an alternative course of metabolism, harmless and usually congenital and lifelong." He refers to it as a "peculiarity." By this point it is clear that his paper is of purely academic interest. The only palliation it would bring would be reassurance to a tiny number of mothers with black-nappied babies.[32]

Alkaptonuria was, however, a fascinating problem, particularly for its involvement with heredity. Garrod shows, once again, that it is "congenital"—a term that combines inherited traits and traits acquired at birth, although he refers only to heredity. The condition tends to run in families, and in his survey a large fraction of known cases occurred in the children of "consanguineous" or cousin marriages. Other authors had discussed this fact, Garrod continues, "but seldom in a strictly scientific spirit. Those who have written on the subject have too often aimed at demonstrating the deleterious results of such unions on the one hand, or their harmlessness on the other, questions, which do not concern us here at all." He is not interested in debating the morality of cousin marriages. Nor was there any reason to think that marrying your relative in and of itself makes you have diseased children. Rather, there must be some "peculiarity of the parents, which can remain latent for generations, but which has the best chance of asserting itself" in the offspring of two related family members. Acknowledging Bateson's assistance, he points out that the law of heredity discovered by Gregor Mendel explains the inheritance of alkaptonuria.[33]

Garrod then puts the pieces together. Combining the known biochemistry with his new genetic data, he refutes the germ theory of alkaptonuria. The metabolism of amino acids such as tyrosine involves many biochemical steps. The black pigment, homogentisic acid, is in fact made in normal human cells, but it lasts only an instant, until it is altered in the next step of the pathway. Alkaptonurics, Garrod reasons, lack the determiner for one step in that pathway. The black pigment fails to be converted, builds up, and is excreted in the urine. Alkaptonuria, then, is a "sport"—what we now call a mutation. Garrod had no way of knowing how many people carried an invisible single copy of the sport—heterozygotes, with one copy of the determiner and no symptoms. But if two carriers marry, a quarter of their children on average will show the trait. This, of course, would happen much more often in a cousin marriage. This—and not some celestial punishment for marrying your cousin—explained the high incidence among consanguineous marriages. Garrod was pleased to strip morality from diagnosis. "Such an explanation," he writes, "removes the question altogether out of the range of prejudice."[34]

At the paper's core is a remarkable leap of intuition and empathy. Garrod extends the idea of the "freak of metabolism"—not in the direction of more serious and common diseases, but toward even rarer, more benign "sports." Although alkaptonuria is conspicuous, he reasons, the loss of other determiners might not be. Other alternative courses of metabolism might be subtler, perhaps even undetectable. If so, Garrod goes on, "the thought naturally presents itself that these are merely extreme examples of chemical behavior which are probably everywhere present in minor degrees." Just as no two individuals are exactly alike physically—a point Darwin had made with exquisite thoroughness—so it may be, thought Garrod, that no two individuals are exactly alike chemically. We are all, he said, chemically individual. His interest was human uniqueness and how it would influence holistic turn-of-the-century medical care.[35]

Garrod found three other conditions—pentosuria, cystinuria, and albinism—that seemed to fit a similar pattern. They all appeared to be alternative courses of metabolism that followed simple Mendelian patterns. (He was right about pentosuria, but the other two turn out to be more complicated, both genetically and biochemically, than he realized.) These four diseases, he felt, went far toward establishing biochemical variation as a frequent if not universal principle of medicine.

He interpreted these results as heralding the biochemical study of human constitution. Individuality could account, he wrote, for differences in sensitivity to different drugs and natural immunity against infecting organisms, and the variations in metabolism that surely underlay obesity, as well as pigmentation differences in skin, hair, and eyes. Essential to Garrod's vision is the idea that there is no "normal." Variation itself is the norm. If one considers the entire constitution, we are all deviants.[36]

Garrod grouped alkaptonuria, his freak of metabolism, along with albinism, pentosuria, and cystinuria under the more dignified name of "inborn errors of metabolism." In 1908 he gave that phrase as the title of the distinguished Croonian lectures, which he delivered before the Royal College of Physicians. With hindsight and the knowledge of their legacy, they seem magisterial. Reviewers at the time responded with cool respect, yawns, peevish jabs, and incomprehension. The obscurity and seeming irrelevance of his subject, combined with a general ignorance of Mendelism among physicians, combined to bury Garrod's masterwork. The *Lancet* published the articles, and the publisher H. Frowde and Hodder & Stoughton collected them into a book the following year, but for decades they received few citations by either scientists or physicians. In 1989 Barton Childs and Charles Scriver found a total of seven references to Garrod out of twenty biochemistry textbooks published between 1900 and 1945.[37]

Garrod never learned much more genetics than what Bateson had taught him—his biographer, the human-geneticist Alexander Bearn, called him a "reluctant geneticist." For that matter, Bateson never learned much more genetics either. Some of Bateson's fire died in 1906, along with Weldon and the end of their feud. He achieved more material success: finally, he got a full-time faculty appointment, as director of the John Innes Horticultural Institute. But his science, drained of passion, fizzled. Having advanced the vanguard, he dug in while the rest of the troops marched past. By 1903 the idea was forming that the Mendelian elements were associated with the chromosomes. By 1910 most scientists accepted that genes were physical objects arrayed along the chromosomes. That year, Thomas Hunt Morgan, a Columbia University biologist, discovered a white-eyed fruit fly in his colony of normal red-eyed flies. More mutations appeared, and, as Morgan and his students bred the mutants, the appearance of new mutations accelerated. Within three years, their breeding program was producing more mutations than they could study. The historian of science Robert Kohler called it the "breeder reactor." Morgan's group, joined by a rapidly expanding community of fruit fly geneticists, mapped the genes to specific spots on the chromosomes. Other researchers mapped genes in corn and other plants as well as mice. The decade of "Mendelism" had passed; genetics, we now say, had entered its "classical" period.[38]

Though Bateson's terminology remained a part of the permanent core of the new science, some of his prettiest ideas crumbled under the weight of data. His presence-and-absence theory fell quickly, as many genes turned out to have several alleles. Morgan's fly boys found more than fifty genes that influenced eye color, disproving (though not eliminating) the unit character idea. Proud, idiosyncratic, stubborn Bateson found the mapping work barbarous, divorced from nature. He refused to relinquish his pure, abstract notion of genes. He would not accept the chromosome theory. He held out until 1921, less than five years before his death. Visiting North America for a lecture, he stopped to visit Morgan's laboratory in New York. He studied the data. He looked through microscopes. He could no longer deny the overwhelming evidence. Genes were real. On his conversion, he wrote to his wife and longtime collaborator, Beatrice, "I was drifting into an untenable position which would soon have become ridiculous."[39]

While Bateson slid from the vanguard to the old guard, Garrod left genetics altogether. He did not keep up with the science through its explosive period in the teens and twenties. He sought neither to explain his data more precisely nor to discover other patterns of inheritance. Nor did he build up a catalogue of inborn errors as many medical researchers might have done. Rather, he spent the rest of his career trying to understand better his notion of chemical individuality.

Garrod saw chemical individuality as a more precise version of the old-fashioned idea of diathesis. The Hippocratic physicians of ancient Greece used that word to describe a disposition—the many ways that one's temperament and constitution affect health. By the early nineteenth century, however, it came to signify a *pre*disposition, the condition of the body that makes it susceptible to certain kinds of diseases. French physicians at this time believed that some diatheses were inherited, such as the herpetic (predisposing to herpes) and gouty diatheses. Others, such as the syphilitic, verminous, and gangrenous diatheses, were acquired. By Garrod's day, diathesis signified an innate predisposition to a particular disease. In that age of epidemics, belief in a tubercular diathesis was so strong that many physicians rejected the notion that the disease was caused by an external germ. If it were, they argued, then why do some people who are exposed not catch the disease? When the Munich physician Max von Pettenkofer quaffed a beaker of virulent cholera bacilli, he aimed to prove that the "seed" of disease can take root only when it falls on fertile "soil."

Yet diathesis and the germ theory were not logically contradictory. Some doctors reconciled them by using a metaphor of seed and soil. The cholera germ, for example, was a seed. It could cause disease, but only if it fell on fertile soil—a body with the right diathesis. Some people were simply "sterile" or unsusceptible to the cholera germ; others were susceptible to different degrees or under different conditions. Yet diathesis, and the related notion of constitution, were empirical observations with no underlying mechanism and no great insights for therapeutics. In contrast, the one germ–one disease theory had clear, immediate implications. Pettenkofer notwithstanding, most people exposed to cholera develop cholera; reducing exposure reduces incidence. In the early twentieth century, then, the clean, modernist lines of the germ theory made diathesis seem overstuffed and musty.[40]

Garrod stood firm against germ-theory hype or dogmatism. "The two factors, the soil and the seed, the external and the internal, must both be taken into account if we are to form any adequate conception of the nature and causation of any disease," Garrod wrote in his 1931 book, *The Inborn Factors in Disease.* Garrod aficionados consider it the culmination of his life's work, the realization of a philosophy of constitutional uniqueness. In it, Garrod developed an elegant, Janus-headed vision of the chemical complexity of life. He looked back to the concept of diathesis and forward to a biochemical explanation of hereditary variation. In recent years, he implied, seeds had received more attention than soil. The soil, for Garrod, was purely chemical. "What our fathers called diathesis," he said, "is nothing else but chemical individuality."[41]

Heredity, of course, was central to individuality, and he acknowledged in the book that traits must have "a molecular representation" in the sperm and eggs. But the *Inborn Factors* is mainly about complexity, not genetics. Complexity is not a by-product of living systems, he said. It is part of the fabric of life. For Garrod, diathesis was a way of grasping the complexity and diversity of life and of health. It described human constitutions the way naturalists described forests, deserts, and grasslands, each with its own character, none more "normal" than another. I suspect this was a tacit point of contact between him and Bateson; the reflective physician is a sort of naturalist of the human body. Garrod even employed natural-history metaphors. "Just as in a mushroom bed a hidden, underground mycelium throws up mushrooms here and there, and from time to time," he wrote, so diathesis is revealed when cases of disease occur. Nature's complexity, full of checks and balances, buffers the organism against infection, stress, and environmental change. It gives stability to the condition of health. In 1932 Walter Cannon called this idea *homeostasis.*[42]

And as Garrod said diathesis was like a mycelium, so too Garrod's ideas. The *Inborn Factors* went underground immediately. It received only a handful of citations in the literature and was mentioned in only a few textbooks. But although no one read his books, Garrod was never forgotten to the extent Mendel was. He was a famous physician, a fellow of the Royal Society and of the Royal College of Physicians, and a Knight Commander of the Order of St. Michael and St. George, one of the highest ranks of British knighthood. He capped his career as Oxford's Regius Professor of Medicine, succeeding the legendary William Osler in 1920. Garrod was not forgotten, but, for a time, his idea was.

A few mushrooms popped up here and there. By 1911 Charles Davenport, dean of American eugenics, had begun citing Garrod's work in his own writing on the inheritance of constitution (see chapter 2). Garrod became something of a cult figure among progressive English geneticists in the nineteen thirties, forties, and fifties. In 1933, for example, Lancelot Hogben discussed Garrod's alkaptonuria work in his minor classic *Nature and Nurture.* He called Garrod's 1902 paper "a landmark in the history of human genetics." In 1950 the *Drosophila* geneticist, Nobel laureate, and antimutation crusader Hermann Joseph Muller referred obliquely to Garrod in "Our Load of Mutations," his landmark essay for the new *American Journal of Human Genetics,* when he mentioned "hereditary abnormalities of metabolism" such as alkaptonuria and phenylketonuria. In 1952, when human genetics was on the cusp of becoming an important field again, the great mathematical geneticist J. B. S. Haldane published a remarkably forward-looking book called *The Biochemistry of Genetics.* Criticizing recent work in the biochemistry of metabolism, he wrote, with typical wit, "If readers turn to Garrod's (1909) 'Inborn errors of metabolism' they will find a much more modern point of view." There is, in short, a thin stream of Garrodian thought trickling straight through to the Galton-Garrod Society.[43]

* * *

Childs, McKusick, Glass, and Lilienfeld named their journal club to express the interplay of Garrodian and Galtonian approaches to heredity and health. Each member brought to the group a distinctive perspective on medical genetics. Childs, the pediatrician, saw many cases of hereditary disease. McKusick studied internal medicine; he had recently described a curious hereditary syndrome of polyps and melanin spots that seemed to indicate an inborn error of metabolism. Glass was a *Drosophila* geneticist with a social conscience. He was drawn to human genetics out of a desire to understand and benefit humankind. And Lilienfeld was an epidemiologist, interested in the manifestation and spread of disease across populations. Drawn together by a shared interest in genetics and human disease, they embodied approaches ranging from clinical, Garrodian biochemical individuality to scientific Galtonian population studies. I suggest that the hybridized name of the club was not a compromise but a nexus—a point of confluence for the members' disparate interests and an expression of a shared vision of genetics.

By invoking Garrod, the society members sought to instate a neglected figure in the biomedical pantheon. Garrod was a well-kept secret, admired by a few cognoscenti—the Haldanes, the Hogbens—and was simply off the intellectual landscape for most biomedical researchers, especially in America. In part as a result of Childs's and McKusick's advocacy, in the 1950s Garrod gained a small but devoted following who considered him the patron of biochemical genetics. The "Garrod" they emulated—the reputation, rather than the man—represented efforts to discover the molecules, the chemical pathways and transformations, the physical substrates of hereditary disease. This required subverting one of Garrod's most cherished and idiosyncratic principles: the harmlessness of most biochemical variation. He liked alkaptonuria as a case study precisely because it seemed to have no serious effects. (This turns out not to be strictly true; alkaptonurics do have some problems later in life, most seriously arthritis.) Individuality per se fascinated him. Like his tutor Bateson, Garrod saw diversity as a biological virtue, a necessity for adapting to a changing environment. There could be no ideal constitution, no state of biochemical purity to strive for. There was only difference; some adaptive, some not, most neutral. Childs was the only one of the quartet to fully embrace this view as a philosophical tenet. The others, to varying degrees, tended to interpret Garrodianism more pragmatically. They sought biochemical explanations of clinically relevant diseases. As we shall see in chapter 7, the study of inborn errors of metabolism grew into a productive and therapeutically important branch of medical genetics. Garrod was not opposed to clinical benefit, of course, but he was more of a philosopher than a craftsman. He searched and researched the constituents of life, diversity, and humanity. Such people are rare. It is understandable that as Garrod became "Garrod," the more ethereal aspects of his thought gave way to the principles that seemed to have more practical benefit.

For the Galton-Garrod group, then, "Garrod" stood for individuality, biochemistry, the search for physiological mechanisms, and the medical side of medical genetics. A Garrodian was comparatively uninterested in panoramic social problems. As Childs put it to me once, he and Garrod were both too interested in patients to be interested in people. Interest in individuality and constitution also implied consideration of the interaction of heredity and environment. "Garrod" symbolized an interest in the constellation of innate and happenstance factors that led to a particular patient having a given disease at that moment in time and in his or her life.

Invoking Galton, in contrast, was an act of rehabilitation. The group had to overlook the fact that the father of population genetics was also the founder of eugenics. In 1950 the memory of the Nazis was still fresh. The Germans had perverted Galton's mildly chauvinistic principles of "racial" improvement into xenophobic policies of racial extermination. By ignoring this fact, the society's members went beyond normal scientific gentility; there was an element of blinkered self-deception in allying themselves with Galton. The quartet had to isolate the scientific from the social, to leave politics outside the laboratory. They had to neglect explicitly eugenic attitudes and practices throughout the history of their respective specialties. Such revisionism is, of course, common practice in science, and it too is understandable and, to an extent, justified. Only an armchair critic can take account of all contexts at once; the alternative is professional paralysis. Still, fortifying the boundary between the laboratory and rest of the world is hazardous—in direct proportion to the technical power and human consequences of the science. Both have increased dramatically for medical genetics in the sixty-plus years since the group was founded.

With Galton's messy, morally freighted social project tabled, the society used *Galton* to stand for a populational approach, the search for patterns in the data, the trends without which clinical case studies become just so many stamps in the collector's album. Which cases, which symptoms are outliers, and which reflect the true causation? What is the signal amid the noise? *Galton* also stood for quantification, putting numbers on the data, and deriving algorithms—"abstracting truth from numbers," as the geneticist Walter Nance once told me. Galtonian approaches are technical, analytic, experimental. And socially, they tend to be more public health oriented, concerned with saving or improving the lives of as many as possible, though this may require neglecting or sacrificing the interests of some individuals. A Galtonian approach, then, enables a broader interest in using science to solve social problems.[44]

Before 1950 it would have been all but inconceivable to identify Galton and Garrod as the emblems of their art. Afterward, it became increasingly inevitable. But the biomedical moment in which the Galton-Garrod Society was formed is an inflection point, not a threshold. Neither then nor at any other time did medical genetics break suddenly with its eugenic past. There is no Galtonian saltation, no intellectual or professional "sport" that became modern medical genetics. Rather, driven by the twin aims of relief of suffering and human improvement, in dynamic tension, each stimulating, constraining, and shaping the other, human heredity evolved gradually from a fringe and disreputable specialty into the heart of biomedicine.

2
Fisher's Quest

JUST AFTER CHRISTMAS, 1904, Irving Fisher kissed his wife, Margaret, and boarded a westbound train out of New Haven, Connecticut. Fisher, a professor at Yale and one of the nation's leading economists, first stopped in Chicago for the American Economic Association annual meeting. He then traveled northeast to Battle Creek, Michigan, home of John Harvey Kellogg's famous Sanitarium. He framed the trip in mythic terms. "I want to fulfill my mission, if it be a mission, for you," he wrote to Margaret from the train. "I am on a quest—not like Ponce de Leon for the fountain of youth but for ideas which may help us to lengthen [life] and to enjoy youth and the spirit of youth."[1]

He set out seeking treatment for tuberculosis, known by then to be caused by a germ, in an institution steeped in the traditions of nineteenth-century alternative medicine and charismatic religion. But in Battle Creek he absorbed the tenets and philosophy of eugenics, a science that claimed to cleanse and purify the germ plasm, as personal hygiene cleansed the body. He became a leading figure in the Progressive-era health reform movement and early attempts to nationalize healthcare. Two decades later, Fisher found himself the charter president of the American Eugenics Society, the nation's leading education and propaganda organization for human hereditary improvement. Though Fisher was neither a physician nor a geneticist, his journey is a foundational, forgotten story in the history of how medicine became genetic.

In the Progressive era, preventive medicine was the health profession's best tool for relief of suffering, and "race betterment" was its holistic approach to human improvement. Heredity connected them seamlessly. The roots of medical genetics lie in a flexible, constitutional approach to heredity and health, one in which prevention and predisposition were at least as important as hard genetic determination, and spirituality was as important as science. The better-known narrative of American human genetics as strictly Mendelian and agricultural is Galtonian, born out of a reversal of Darwinian logic in which the artificial selection used by animal and plant breeders was applied to human beings. Its complement is a more Garrodian medical approach, which attempted to tease out the threads of constitution with an eye toward prevention of disease (both hereditary and not). Fisher's quest is the story of that Garrodian approach and how it interwove with the better-known Galtonian style of eugenics; the two were not opposed but complementary.[2]

* * *

Born in 1867 in Saugerties, New York, Fisher was a child of god and the son of a Congregational Christian minister. He was highly disciplined, almost totally humorless, and denied himself caffeine, alcohol, tobacco, and pepper. He was a devoted father and a romantic husband: when he traveled, he wrote adoring, sentimental letters to his wife every day. He spent his entire intellectual life at Yale University. As a student, he took courses in electricity, magnetism, and thermodynamics with the physicist Willard Gibbs, and in political economy, finance, and politics with the economist and sociologist William Graham Sumner. Fisher remained cordial with, even admiring of, Sumner, but he rejected Sumner's gloomy reverse Darwinism, in which humanitarian aid and social regulation were seen as brakes on human progress; Fisher preferred instead a collectivist, bureaucratic vision of mutual aid and paternalism. In 1891 he completed his Ph.D. in mathematics and was appointed assistant professor in that department. Four years later, he transferred to the Department of Political Economy, where he spent the rest of his academic career, until his retirement in 1935. He wrote some twenty books, including *The Nature of Capital and Income, The Rate of Interest, The Purchasing Power of Money, The Theory of Interest, Booms and Depressions*, and three books on Prohibition. He campaigned for Prohibition and the League of Nations, headed the American Economic Association, and founded the Econometric Society. He earned a fortune in the 1920s and lost it in the first years of the Depression. His private writings indicate a strongly deterministic outlook rooted in religious faith, and an irrepressible optimism for the progress of society.[3]

Economics was Fisher's profession, but health was his calling. In 1898 he contracted tuberculosis; his five-year battle with the disease transformed him into a health crusader. The rigor of sitting in a raccoon coat in the middle of an upstate New York winter, writing to Margaret from the porch of Edward Trudeau's sanatorium—in pencil because ink would freeze—stirred in him a passion for virtue as vigor. He worshiped at the temple of the body, with an evangelical belief that purity of diet, regimen, and spirit were the keys to a long, healthy, and happy life. Although he would approach the subject scientifically, his motivation was spiritual. Having felt and conquered TB's terrifying restriction of breath, Fisher came to view the absence of disease as perfection, and normalcy as a form of enlightenment. "I have developed a passion for out-of-door living," he wrote in 1903, from Trudeau's sanatorium. "Last night at sunset I sat out here like an Indian, thinking of nothing, but *feeling* the serenity and power of the Universe. The joy of living and breathing is joy when living and breathing are normal."[4]

This association between piety and public health has deep roots in American culture. During the middle third of the nineteenth century, such health-oriented postmillenarian religious cults as the Seventh Day Adventists emerged, as well as such spiritualist medical cults as homeopathy, osteopathy, and hydropathy. Health, healing, purity, and spirituality became highly fungible. Both sets of sects aimed toward purification—they shared a kind of physical Puritanism, a worship of the body uncontaminated by chemical poisons or toxic thoughts and feelings. Disease could be seen as a kind of sin. Clean body, clean mind, clean spirit—each depended on the others. Purification is an effort to reach an ideal, uncontaminated state; there was a strain of perfectionism to the enterprise. For example, the "water cure," of which Charles Darwin was a devotee, was partly a hygienic act of purification. And in the 1830s the New York public health reformer Robert Hartley was instrumental in transforming the concept of "temperance" in alcohol use from its literal denotation of moderation to a connotation of total abstinence. For Hartley, purification admitted of no half-measures.[5]

Thus, heredity, piety, and health are an old and durable trio with a broad repertory, well beyond the American border. Germany in the 1890s had what Sheila Faith Weiss called a "veritable fetishism of heredity," which found expression in numerous neurological and psychiatric studies encompassing such conditions as epilepsy, hysteria, criminality, and feeblemindedness. German doctors had come to feel like stewards of the national health. Many pushed for better hygiene as a path to health; those who also considered posterity attended as well to what came to be called "race hygiene." In France, too, the hereditarian tradition had for decades been grounded in a logic of social control. Thus, from the nineteenth century, European conceptions of heredity and health tended to be technocratic; an approach that entered the American dialogue later and incompletely.[6]

Allopathic or conventional medicine, of course, was also invested in concepts of heredity. In chapter 1, we saw that orthodox physicians in the last quarter of the nineteenth century drew on medicine's ancient interest in heredity as a bulwark against the reductionist zeal of the germ theory fanatics. Passionate followers of Pasteur and Koch seemed to claim that germs were not only necessary but sufficient to cause infectious disease. Heredity, their more mainstream colleagues countered, was one of several factors contributing to predisposition or diathesis, the tendency to be sensitive to a given disease or class of illness. Doctors had long appreciated the importance of heredity in disease. But at the turn of the twentieth century, only those few who embraced the eugenic fervor could see any relevance of Mendelism's mathematical formalities for their daily practice. Physicians therefore had a complex, multivocal response to eugenics as it took hold in the new century. While some resisted it as heartless and antimedical, others embraced it as a form of preventive medicine.

* * *

The Western Health Reform Institute at Battle Creek, Michigan, was the epitome of physical puritanism. Founded in 1863, fifty miles west of Ann Arbor, Battle Creek was the home of the Seventh Day Adventists, a postmillenarian Christian denomination distinguished by their observation of

Saturday as the Sabbath. Adventists believed in vegetarianism and the preservation of health in all ways, "tho now-a-days," Fisher wrote to Margaret, "they have fallen from their original strictures." "This," he continued, "is how Dr. K. evidently got started on his ideas." Indeed, "Dr. K."—John Harvey Kellogg—was nurtured by Adventists. The Kelloggs were a founding family of Battle Creek and were close to Ellen White, considered the prophet of Seventh Day Adventism. John Harvey, born in 1852, spent many hours in Ellen White's home. As a child, he was a precocious reader and highly musical. At his brother's request, he attended the Hygieo-Therapeutic Institute of Russell T. Trall in New Jersey. Kellogg decided to become a health reform doctor, turned toward medicine, and took an M.D. from New York's Bellevue Hospital medical school. In 1876 he assumed the directorship of the institute.[7]

Kellogg combined a Protestant work ethic, a preacher's charisma, a doctor's compassion, and a scientist's discipline. He developed a comprehensive program for living, complete with a vegetarian diet that included the introduction of cold cereals, such as flaked corn and wheat meal, for breakfast. (In a bitter internecine legal battle, his younger brother, William Keith, won the patents and the famous cereal company.) He also believed in hydrotherapy, massage, exercise, plying the digestive tract from both ends with active culture yogurt, and attentive, vigorous medical monitoring. He rejected drugs of any kind as poisons. Kellogg created not a mere medical facility but an entire way of life, with full menus, purpose-built architecture and machinery, and a strict daily regimen. It was a seductive mixture of pageantry, intimacy, evangelism, and technology. Kellogg modified the older term *sanatorium* and renamed the institute the Sanitarium, to connote the holistic way in which he not merely treated disease but actively increased health. Follow me, Kellogg said, and you will be healthy in body, mind, and spirit. He called this holistic approach "biologic living."

In February 1902 a fire destroyed "the San"; Fisher arrived at the newly rebuilt facility, Kellogg's ultimate monument to hygiene, heredity, and health. Fisher, initially skeptical of the hype that already surrounded Kellogg and Battle Creek, had to admit that both were impressive. The facility was "magnificent," he wrote to Margaret, "substantial and fireproof. Six stories 200 paces long looks like Hotel California . . . only three times as large."

Spread over some four hundred acres, the San could accommodate seven hundred guests, tending to them with thirty physicians and a staff of up to one thousand in peak months. The accommodations were luxurious. "My room is large private bath gilded iron bedstead, steam heat, electric light, telephone, special ventilating arrangements and very good ones." The director also impressed him greatly. "Dr. Kellogg too surpasses my expectations," he wrote. "He is short . . . but quick energetic and a terrific worker. This I know even if I need proof of his assertion that he has often worked 36 hrs. on a stretch without food or sleep and could work 48 if need be!" Fisher remained mum to Margaret on the cuisine; perhaps a menu he pocketed bears mute testimony as to why. A diet of sanitas steak and tomato, nut cutlets, bromose, potted protose, and "no-coffee" with cream and sugar was apparently nothing to write home about.[8]

Heredity was an important dimension of biologic living. Kellogg's conception of heredity was simple-minded by the standards of late-nineteenth-century thought—it was a scientistic folk heredity, rooted in the inheritance of acquired characteristics associated with Jean-Baptiste Lamarck, rather than in Darwinian natural selection. On the one hand, Kellogg believed that better hygiene alone would lead to the deterioration of the race by

2.1 The opening, in 1903, of the rebuilt Battle Creek Sanitarium—starting point of Fisher's quest. From Kellogg, *Battle Creek Sanitarium System*, 18

propagating the weak and feeble; on this, he was perfectly in tune with the eugenic thinkers who emerged in the Progressive era. On the other hand, he believed that immorality created bad heredity, while biologic living improved the germ plasm. In *Plain Facts for Old and Young* (1880), Kellogg expressed a repressed sexuality that Progressive eugenicists three decades hence would have found entirely congenial: "In no other direction are the effects of heredity to be more distinctly traced than in the transmission of sensual propensities. The children of libertines are almost certain to be rakes and prostitutes." Such Lamarckian views blurred the distinction between heredity and environment; in line with the worst eugenic fallacies, Kellogg thought that anything that ran in families was self-evidently hereditary.[9]

Smitten with Kellogg's personality, method, and philosophy, Fisher returned to Yale to begin a series of experiments with the physiologist Russell Chittenden on college athletes. He put strapping male Elis on various vegetarian diets, put them on dynamometers, measured their endurance and its relationship to constitution, caloric intake, and nutritional quality. During this period, he began to evangelize Kellogg's gospel and vocabulary. Crossing campus one day, he saw then–Secretary of War (and Yale grad) William Howard Taft. "I buttonholed him and after steering into the subject asked him why he didn't try the Chittenden [nutritional] scheme in the army. He himself is evidently very high proteid." Fisher, steeped in Kellogg's idiosyncratic jargon, could hardly believe that an educated person would be ignorant of it. "In the course of the talk he asked, 'What is proteid?'!" Fisher, a distinguished professor and a community leader, took an active role in his local antituberculosis association. He began to read the scientific literature, to give public talks—to become, in short, a health crusader.[10]

* * *

American human genetics is often portrayed as exceptional in its agricultural origins, in contrast to more medically oriented human genetics in France, Germany, and Scandinavia. In the United States, the advent of Mendelism appealed to breeders immediately as a set of tools for improving crops and livestock. Consequently, the first professional society of geneticists was the American Breeders' Association, which met for the first time in

St. Louis at the end of 1903, under the auspices of the Association of Agricultural Colleges and Experiment Stations. Willet M. Hays, a plant breeder at the University of Minnesota, opened the proceedings by asking practical breeders to "occasionally pause and study the laws of breeding" alongside academic students of heredity. "The wonderful potencies in what we are wont to call heredity," he wrote, "may in greater part be placed under the control and direction of man." The accelerating growth of the land grant universities after 1862 provided a natural home for those interested in hereditary improvement, whether of plants, animals, or humans. In 1906, when the formal structure of the ABA was established, it included a committee on eugenics. Chaired by David Starr Jordan, president of Stanford University, the committee included many pioneers of American eugenics, such as Davenport, Alexander Graham Bell, and Roswell Johnson.[11]

These agricultural ties give eugenics a horrible fascination for us. A popular trope of the day—dating back to Francis Galton—was to decry the fact that so much effort was put into breeding horses, so little into breeding humans. This agricultural metaphor added potency to old language about the human "stock." Such objectification of humans is deeply offensive to our ears today. Yet the offense can be overstated. The historian Martin Pernick has pointed out that in reference to humans, *breeding* has traditionally meant both heredity and upbringing—someone who is "well-bred" was not only born well but reared well. Thus when Charles Davenport wrote in 1915 that eugenics was the "science of improving mankind through better breeding," it was understood that heredity and environment both contributed.[12]

Charles Benedict Davenport is known among historians as an arch-hereditarian, an intelligent man who became so blinded by the simplistic power of Mendelism that he proposed a single-gene trait he called "thalassophilia," or love of the sea, which he said could account for lineages of sea captains. But he was also an accomplished geneticist, and he published and lectured extensively on genetics and medicine. His extensive studies of human heredity are a bag of scientific allsorts—among frivolous and blinkered projects are a few pioneering classics. He introduced methods that have shaped the study of the heredity of disease for more than a century. He needs to be taken seriously, for both good and ill, as a founding figure of medical genetics.[13]

A Connecticut-bred, Harvard-trained zoologist, the young Davenport was a pioneer of marine ecology. In 1898 he accepted a post on the other side of Long Island Sound and became director of the Biological Laboratory at Cold Spring Harbor, founded in 1890 as New York's answer to the Marine Biological Laboratory at Woods Hole, Massachusetts. The "Bi Lab" operated summers only, offering "tables" to visiting scientists, in the tradition of such European marine stations as the Stazione Zoologica Naples, as well as nature study courses for local children. As the new director, Davenport

2.2 Charles Benedict Davenport, c. 1903. Courtesy of American Philosophical Society Library

began to turn the Bi Lab into a serious scientific research and teaching institution. Within five years he established as prerequisites for entrance into the summer courses both a college degree and a serious interest in original research.[14]

Davenport was among the first scientists in North America to embrace Mendelism, especially its connection to Darwinian evolution. In 1902 the industrialist Andrew Carnegie gave $10 million to endow the Carnegie Institution of Washington, a foundation to strengthen American education and research in science. Twelve days after the Carnegie Institution was incorporated, Davenport presented its board of directors a plan to establish an endowed or fully supported laboratory that would be devoted to experimental studies of evolution, including by Mendelian hybridization and de Vriesian mutation. In the late summer of 1902 he and his wife Gertrude (also an accomplished geneticist) toured the marine laboratories of Europe and arranged to visit Francis Galton and his disciple Karl Pearson at the Galton Laboratory in London. Impressed by Davenport's projected plan for a concerted attack on the major problems of heredity and evolution, both Galton and Pearson wrote letters of recommendation to the Carnegie on his behalf. "Personally he seems to me stronger than his published work," Pearson wrote. "I should say he would not be wanting in energy and keenness of interest and would keep himself in touch with European workers and methods."[15]

Davenport's institution would combine elements of academic studies of heredity and the kinds of experimental programs being worked out at agricultural experiment stations throughout the East and Midwest. It would be more academic in orientation than, say, the Connecticut Agricultural Experiment Station across Long Island Sound, but it would nevertheless have much of the same character. There was some discussion about the name of the Carnegie Station at Cold Spring Harbor. Billings suggested calling it simply the Biological Laboratory. Davenport disagreed strongly, feeling, probably rightly, that adding yet another "Biological Laboratory" to the growing list that included the other Cold Spring Harbor Laboratory, the Marine Biological Laboratory at Woods Hole, and others, would add little distinction to the Carnegie Institution's bold new enterprise. Rather, he said, call it the Station for Experimental Evolution of the Carnegie Institution.

So it was called when it opened on January 19, 1904, although Billings quipped, "While the Carnegie Institution may be in need of experimental evolution, I do not think that this institution is established for the purpose of evoluting it."[16]

From experimental evolution it was a short hop to eugenics. The Davenports began publishing papers on the inheritance of such human traits as hair and eye color. Davenport became involved with the ABA Eugenics Section. And he was seeking a third institute for his empire, one that would focus on problems of human heredity, both applied and fundamental. Consulting *Who's Who* for wealthy but deceased Long Islanders, he discovered that recently the railroad magnate E. H. Harriman had died and left his fortune to his wife. When Davenport paid the widow Harriman a visit, he exploited the medical applications of eugenics. On one arm he had the eminent Johns Hopkins physician William H. Welch, a strong advocate of eugenics as a means of understanding and preventing disease. And on the other he had the celebrated inventor Alexander Graham Bell, whose congenitally deaf wife inspired not only the telephone but an abiding scientific interest in hereditary deafness. Flanked by such celebrities, Davenport persuaded Mrs. Harriman to purchase and donate an estate of seventy-five acres up the hill from the Carnegie Station. The property contained a large but rundown house that at first had no heat or running water. Mrs. Harriman let Davenport fix up the house and improve the grounds, she paid for the construction of a new building on the grounds, and she funded the research. The Eugenics Record Office opened on October 1, 1910. Its program was formulated under the auspices of the ABA. Mrs. Harriman's patronage lasted nearly eight years and totaled nearly half a million dollars, with another $300,000 given for endowment at the end.[17]

Davenport selected Harry H. Laughlin, a schoolteacher and administrator from Iowa, to manage the ERO. Laughlin quickly became the ERO's public face, advising government officials, writing pamphlets, and giving public lectures. He was certainly one of the country's most zealous eugenics advocates, and his fanaticism was a large part of the rise and the eventual decline in the ERO's reputation. Davenport also assembled a Board of Scientific Directors, which reflected the growing prestige of eugenics. The board included Bell as chairman, Fisher, Welch, Lewellys Barker (also of Johns

Hopkins), and the distinguished psychiatrist Elmer E. Southard as members, and Davenport as secretary. Fisher would remain closely associated with these men for the rest of his career. In 1918 the Carnegie Institution assumed control of the ERO, folding it and the Station for Experimental Evolution into a single Department of Genetics. Thus the Carnegie saw itself as supporting both basic and applied genetics.[18]

The ERO introduced a novel and durable method of collecting human genetics data. Elaborating on Galton's idea of the eugenics "record," Davenport developed a questionnaire of the type Galton used. But instead of marching people through his research kiosk one by one, Davenport used mass-mailing, and, most effectively, "fieldworkers" to collect the data. The fieldworkers—more than 250 of them between 1910 and 1924—were mostly young women, many of them nursing students from the New York City area. They were all volunteers—for many of them, working for the ERO was a kind of internship in hereditary public health work. Trainees spent a summer in Cold Spring Harbor, where they received twenty-five lectures encompassing interviewing methods, construction of pedigrees, and the elements of statistics and biometry. The young women then went out into the field for a year, where they catalogued and documented the hereditary patterns of the diseased and insane. The data were recorded on three-by-five-inch cards and stored in a fireproof vault back at Cold Spring Harbor. By 1924, fieldworkers had filled out and filed 750,000 cards. The pedigrees and family histories were analyzed for their fit to a simple Mendelian explanation.[19]

As part of their training, they studied the "Trait Book," a catalogue of humanness, organized in a sort of eugenic Dewey decimal system. Davenport compiled and adapted classificatory schemes from the Census Bureau on occupations, crimes, and causes of death. On mental traits, he consulted the respected—and eugenically minded—psychologists Edward L. Thorndike and Robert M. Yerkes. The result is a chaotically holistic compendium of human characters, covering behavioral traits, biometrical data, and medical conditions chronic, acute, and incidental. Davenport deliberately cast his net as broadly as possible, intending the list to be winnowed with time. Traits in the Trait Book include physical beauty, finger-clasping, breast cancer, malaria, beekeeping, alcoholism, nicotinism, length

2.3 Fieldworker training, Cold Spring Harbor Eugenics Record Office. Courtesy of Cold Spring Harbor Laboratory

of eyebrow hairs, number and quality of skin glands, polydactylism, pinna shape—including descriptions of helix, tragus, lobule, and Darwin's tubercle—Huntington's chorea, migraine, prostitution, nine forms of suicide, imagination, tone-deafness, pyrophobia, philoprogenitiveness, foresight, altruism, frigidity, retinitis pigmentosa, nightblindness, asthma, hemophilia, alkaptonuria, Bright's disease, impotence, and breech birth. The Trait Book flattened and equalized all human characters, from nature to nurture, from subjective to objective, from health to disease, and from trait to habit.[20]

Davenport acknowledged that many of the traits in the book, especially the diseases, were not "hereditary" in the common sense, but he insisted à la Garrod that infectious diseases such as tuberculosis, syphilis, and the plague are products of a "specific germ acting on a susceptible protoplasm" and that that susceptibility could be heritable. Not long ago, the range of traits of interest to these early Mendelian eugenicists seemed a hopeless tangle of heredity and environment, and Davenport a foolish and naïve genetic determinist. For much of the late twentieth century, human-geneticists focused on medical traits with high heritability. But lately researchers have once again become interested in "complex" traits—common, socially relevant, and genetically weak. The majority of the medical and psychological traits in the Trait Book are of interest to biomedicine today, and several have become models of particular genetic or molecular mechanisms. Medical geneticists in 2012, like Davenport a century before, believe that all traits—hereditary, infectious, or experiential—have a genetic component.[21]

Davenport marketed the fieldwork method remarkably effectively. In 1911 he published his masterwork, *Heredity in Relation to Eugenics*, an uneven book that established him simultaneously as the foremost human-geneticist in America and as the overconfident and dogmatic leader of the American eugenics movement. The same year, a thirty-page monograph on the "Study of Human Heredity, Methods of Collecting, Charting and Analyzing Data" appeared, in a new serial publication, *Eugenics Record Office Bulletin*. In 1913 Davenport arranged to send a fieldworker to the psychiatrist Adolf Meyer, at Johns Hopkins. The fieldworker would be used for "eugenical studies" relating to Meyer's psychiatric work. Meyer was so happy with the arrangement that he extended the term for a second six months.[22] In 1917 the New York State Board of Charities adopted the method, publishing an extensive fieldwork manual and crediting the ERO for originating the technique. The manual complicates some of the stereotypes of Progressive-era eugenics. Although explicitly eugenic, it is not dogmatically hereditarian. It takes care to introduce skepticism of the idea of degeneration and to point out that elimination of a putative bad branch of heredity is a matter not to be undertaken lightly. Yet as an example record, it offers the fictional case of "William S. Incorrigible," thus squandering the moral capital of the preceding pages on a cheap joke. Such are the pressures of culture and the risks of wit. But through such means the fieldwork method passed down through the generations like an heirloom, carrying with it some of the commitments and thought styles of the ERO. Fieldwork became a mainstay of twentieth-century medical genetics research.[23]

Davenport, Laughlin, and the ERO have come to symbolize a class-conscious, racially charged, conservative style of eugenics often called "mainline" eugenics. The epitome of that style could be found an hour to the

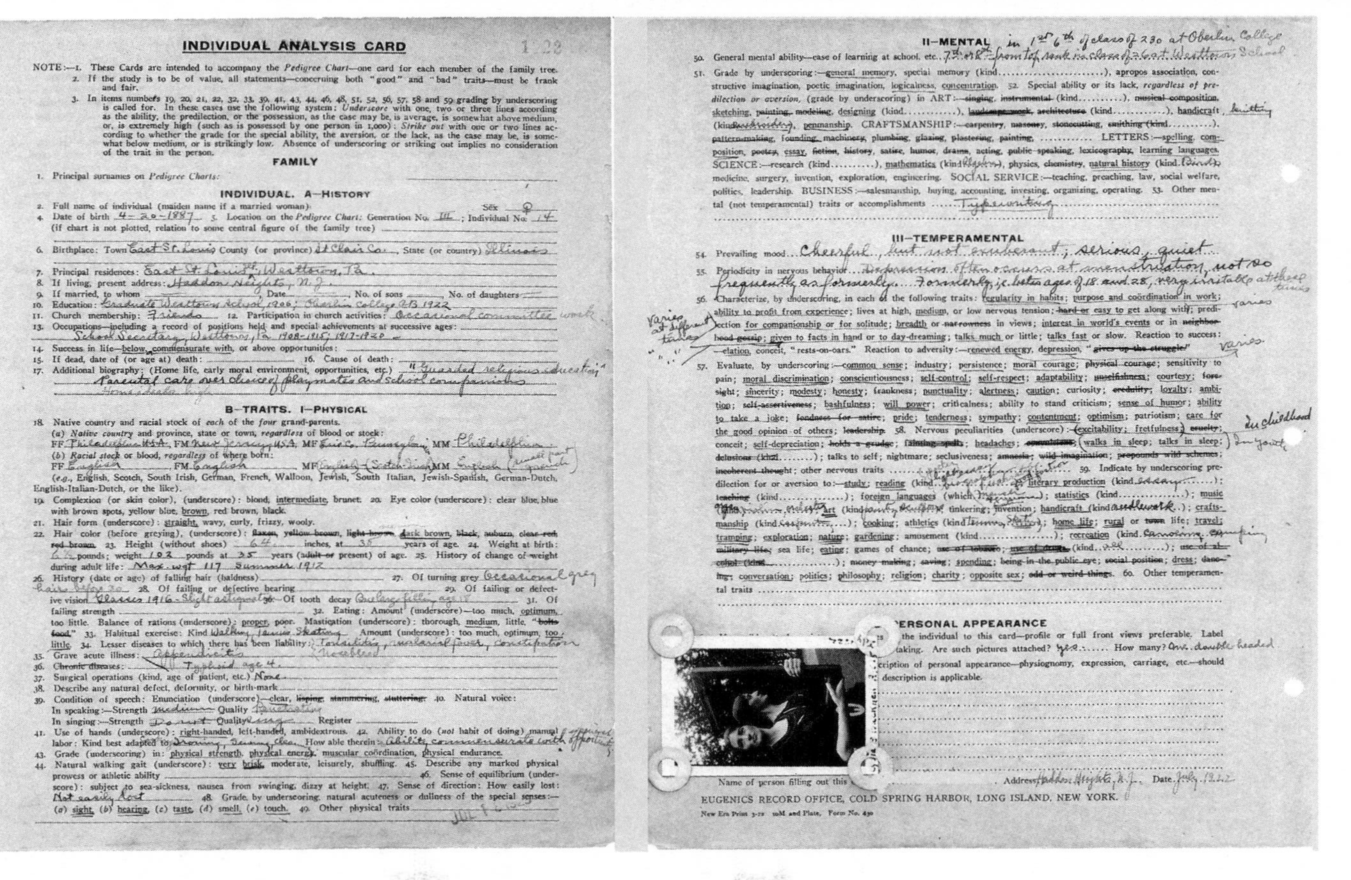

INDIVIDUAL ANALYSIS CARD

NOTE:—1. These Cards are intended to accompany the *Pedigree Chart*—one card for each member of the family tree.
2. If the study is to be of value, all statements—concerning both "good" and "bad" traits—must be frank and fair.
3. In items numbers 19, 20, 21, 22, 32, 33, 39, 41, 43, 44, 46, 48, 51, 52, 56, 57, 58 and 59 grading by underscoring is called for. In these cases use the following system: *Underscore* with one, two or three lines according as the ability, the predilection, or the possession, as the case may be, is average, is somewhat above medium, or, is extremely high (such as is possessed by one person in 1,000): *Strike out* with one or two lines according to whether the grade for the special ability, the aversion, or the lack, as the case may be, is somewhat below medium, or is strikingly low. Absence of underscoring or striking out implies no consideration of the trait in the person.

FAMILY

1. Principal surnames on *Pedigree Charts:*

INDIVIDUAL. A—HISTORY

2. Full name of individual (maiden name if a married woman) Sex ♀
4. Date of birth 4-20-1887 5. Location on the *Pedigree Chart:* Generation No. III; Individual No. 14
(if chart is not plotted, relation to some central figure of the family tree)
6. Birthplace: Town East St. Louis County (or province) St Clair Co., State (or country) Illinois
7. Principal residences: East St. Louis, Westtown, Pa.
8. If living, present address: Haddon Heights, N.J.
9. If married, to whom, Date, No. of sons, No. of daughters
10. Education: Graduate Westtown School, 1906; Oberlin College AB 1922
11. Church membership: Friends 12. Participation in church activities: Occasional committee work
13. Occupations—including a record of positions held and special achievements at successive ages: School Secretary, Westtown, Pa 1908-1915; 1917-1920
14. Success in life—below, commensurate with, or above opportunities:
15. If dead, date of (or age at) death: 16. Cause of death:
17. Additional biography: (Home life, early moral environment, opportunities, etc.) "Guarded religious education" Parental care over choice of playmates and school companions

B—TRAITS. I—PHYSICAL

18. Native country and racial stock of *each* of the *four* grand-parents.
(*a*) *Native country* and province, state or town, *regardless* of blood or stock:
FF Philadelphia USA, FM New Jersey USA, MF Erie Co. Pennsylvania, MM Philadelphia
(*b*) *Racial stock* or blood, *regardless* of where born:
FF English, FM English, MF English (Scotch-Irish), MM English
(*e.g.*, English, Scotch, South Irish, German, French, Walloon, Jewish, South Italian, Jewish-Spanish, German-Dutch, English-Italian-Dutch, or the like).
19. Complexion (or skin color), (underscore): blond, intermediate, brunet. 20. Eye color (underscore): clear blue, blue with brown spots, yellow blue, brown, red brown, black.
21. Hair form (underscore): straight, wavy, curly, frizzy, wooly.
22. Hair color (before greying), (underscore): flaxen, yellow brown, light brown, dark brown, black, auburn, clear red, red brown. 23. Height (without shoes) 64 inches, at 35 years of age. 24. Weight at birth: pounds; weight 103 pounds at 35 years (adult or present) of age. 25. History of change of weight during adult life: Max. wgt 117 Summer 1912
26. History (date or age) of falling hair (baldness) 27. Of turning grey Occasional grey hairs before 30 28. Of failing or defective hearing 29. Of failing or defective vision Glasses 1916 - Slight astigmatism 30. Of tooth decay 31. Of failing strength 32. Eating: Amount (underscore)—too much, optimum, too little. Balance of rations (underscore): proper, poor. Mastication (underscore): thorough, medium, little, "bolts food." 33. Habitual exercise: Kind Walking, Tennis, Skating Amount (underscore): too much, optimum, too little. 34. Lesser diseases to which there has been liability: Tonsilitis, malarial fever, constipation
35. Grave acute illness: Appendicitis
36. Chronic diseases: Typhoid age 4.
37. Surgical operations (kind, age of patient, etc.) None.
38. Describe any natural defect, deformity, or birth-mark
39. Condition of speech: Enunciation (underscore)—clear, lisping, stammering, stuttering. 40. Natural voice:
In speaking:—Strength medium Quality
In singing:—Strength Do not. Quality Register
41. Use of hands (underscore): right-handed, left-handed, ambidextrous. 42. Ability to do (*not* habit of doing) manual labor: Kind best adapted to How able therein:
43. Grade (underscoring) in: physical strength, physical energy, muscular coördination, physical endurance.
44. Natural walking gait (underscore): very brisk, moderate, leisurely, shuffling. 45. Describe any marked physical prowess or athletic ability 46. Sense of equilibrium (underscore): subject to sea-sickness, nausea from swinging, dizzy at height. 47. Sense of direction: How easily lost: Not easily lost 48. Grade, by underscoring, natural acuteness or dullness of the special senses:—(*a*) sight, (*b*) hearing, (*c*) taste, (*d*) smell, (*e*) touch. 49. Other physical traits

II—MENTAL

50. General mental ability—ease of learning at school, etc.
51. Grade by underscoring:—general memory, special memory (kind ...), apropos association, constructive imagination, poetic imagination, logicalness, concentration. 52. Special ability or its lack, *regardless of predilection or aversion,* (grade by underscoring) in ART:—singing, instrumental (kind ...), musical composition, sketching, painting, modeling, designing (kind ...), landscape work, architecture (kind ...), handicraft (kind ...), penmanship. CRAFTSMANSHIP:—carpentry, masonry, stonecutting, smithing (kind ...), pattern-making, founding, machinery, plumbing, glazing, plastering, painting, ... LETTERS:—spelling, composition, poetry, essay, fiction, history, satire, humor, drama, acting, public speaking, lexicography, learning languages. SCIENCE:—research (kind ...), mathematics (kind Algebra), physics, chemistry, natural history (kind ...), medicine, surgery, invention, exploration, engineering. SOCIAL SERVICE:—teaching, preaching, law, social welfare, politics, leadership. BUSINESS:—salesmanship, buying, accounting, investing, organizing, operating. 53. Other mental (not temperamental) traits or accomplishments Typewriting

III—TEMPERAMENTAL

54. Prevailing mood Cheerful, but not exuberant; serious, quiet
55. Periodicity in nervous behavior
56. Characterize, by underscoring, in each of the following traits: regularity in habits; purpose and coördination in work; ability to profit from experience; lives at high, medium, or low nervous tension; hard or easy to get along with; predilection for companionship or for solitude; breadth or narrowness in views; interest in world's events or in neighborhood gossip; given to facts in hand or to day-dreaming; talks much or little; talks fast or slow. Reaction to success:—elation, conceit, "rests-on-oars." Reaction to adversity:—renewed energy, depression, "gives up the struggle."
57. Evaluate, by underscoring:—common sense; industry; persistence; moral courage; physical courage; sensitivity to pain; moral discrimination; conscientiousness; self-control; self-respect; adaptability; unselfishness; courtesy; foresight; sincerity; modesty; honesty; frankness; punctuality; alertness; caution; curiosity; credulity; loyalty; ambition; self-assertiveness; bashfulness; will power; criticalness; ability to stand criticism; sense of humor; ability to take a joke; fondness for satire; pride; tenderness; sympathy; contentment; optimism; patriotism; care for the good opinion of others; leadership. 58. Nervous peculiarities (underscore):—excitability; fretfulness; cruelty; conceit; self-depreciation; holds a grudge; fainting spells; headaches; convulsions; walks in sleep; talks in sleep; delusions (kind ...); talks to self; nightmare; seclusiveness; amnesia; wild imagination; propounds wild schemes; incoherent thought; other nervous traits ... 59. Indicate by underscoring predilection for or aversion to:—study; reading (kind ...); literary production (kind essay); teaching (kind ...); foreign languages (which ...); statistics (kind ...); music (kind ...); art (kind ...); tinkering; invention; handicraft (kind needlework); craftsmanship (kind ...); cooking; athletics (kind Tennis, Skating); home life; rural or town life; travel; tramping; exploration; nature; gardening; amusement (kind ...); recreation (kind ...); military life; sea life; eating; games of chance; use of tobacco; use of drugs (kind ...); use of alcohol (kind ...); money making; saving; spending; being in the public eye; social position; dress; dancing; conversation; politics; philosophy; religion; charity; opposite sex; odd or weird things. 60. Other temperamental traits

PERSONAL APPEARANCE

... the individual to this card—profile or full front views preferable. Label ... taking. Are such pictures attached? yes How many? One, double headed
... cription of personal appearance—physiognomy, expression, carriage, etc.—should ... description is applicable.

Name of person filling out this ... Address Haddon Heights, N.J. Date July 1922

EUGENICS RECORD OFFICE, COLD SPRING HARBOR, LONG ISLAND, NEW YORK.

New Era Print 3-22 10M and Plate, Form No. 430

2.4 Individual Analysis Card, from Fitter Family Study, Eugenics Record Office. Courtesy of American Philosophical Society Library west of Cold Spring Harbor, at the American Museum of Natural History in Manhattan. There, director Henry Fairfield Osborn hosted a kind of eugenic salon he called the Galton Society. Founder Madison Grant, whose 1916 *The Passing of the Great Race* became a touchstone of racialist eugenic thought, intended the Galton Society as an alternative to the American Anthropological Association, dominated at that time by the liberal Franz Boas. In addition to eugenic movers and shakers such as Osborn, Grant, Davenport, and Laughlin, promising young geneticists were invited to attend meetings—men such as Raymond Pearl, an agricultural geneticist who would later head a "Constitutional Clinic" at Johns Hopkins, and Hermann Joseph Muller, a young, extraordinarily bright *Drosophila* geneticist with a passion for practical, humanitarian applications of hereditary knowledge. The Galton Society was all-white, all-male, predominantly Protestant, and mostly wealthy. Their brand of eugenics was heavily scented with leather and cigar smoke, and it focused on such questions as immigration and "the Negro problem." As a primarily anthropological group, the Galton Society was fundamentally Galtonian in approach, concerned with populations rather than individuals. The narrative of agricultural origins, the obsession with race, class, and intelligence, and the overt elitist racism of groups such as the Galton Society create a compelling and horrific story of eugenic objectification.[24]

This narrative is rich with moral lessons, but it neglects the substantial medical and public health dimension of Progressive-era eugenics. This medical strain of American eugenics has gone largely unrecognized. But in America, too, medicine overlapped significantly with concerns over "degeneration"—many of the defects eugenicists sought to eliminate were medical conditions. For example, tuberculosis, the disease that led Fisher to the San, was the subject of intense eugenic investigation. The Belgian physician Albert Govaerts spent the year 1921–22 at Cold Spring Harbor, analyzing ERO statistical data and conducting his own studies. He found that susceptibility to tuberculosis had "indirect" heredity (today we would call it a "complex genetic trait"). The power to resist the disease was inherited as part of an individual's constitutional makeup. This much was mainly a formalization of a long and widely held belief among physicians about resistance to TB. Govaerts sought and found evidence that this predisposition varied by race: American Indians, Negroes, and Hawaiians were unusually susceptible, while Jews and Gypsies seemed relatively immune. Such categories lacked the formal precision of today's genomic haplotype groupings, but they share an essential strategy. The data were still too ambiguous to make very clear-cut recommendations—researchers argued over whether the best way to strengthen the germ plasm was to encourage or discourage marriages between the tuberculous and the nontuberculous. But one point is clear: for these workers, eugenic advice started at the level of the individual and radiated outward to the population. Physicians and patients alike wanted to know whether this particular family would likely have tuberculous offspring. Concern then extended to the community—who should this person marry or not marry?—and thence to the society at large. Hereditary preventive medicine operated on every social level.[25]

* * *

Eugenics tapped the new spirit of Progressive reform. When the ABA Eugenics section and the Carnegie Institution were forming, the democratic, agrarian populism of the 1880s and 1890s was giving way to a more bureaucratic and urban style of reform. Early-twentieth-century reformers reacted against the laissez-faire nineteenth-century attitude that had built the steel mills, the oil industry, the railroads, and the banking system by taking control and regulation as hallmarks of civilized life. Government agencies, private philanthropies, and public charities and activist groups multiplied, each devoted to an urgent social need—society, culture, lifestyle, environment. Shared social languages emerged, of social bonds and the value of scientific efficiency, which focused public efforts on rational collective action. There was a lot to address. Unchecked capitalism had degraded the health and well-being of the working classes for the benefit of the rich. And the influx of workers had led to seismic shifts in population size and character that created slums, cramped and dangerous working conditions, crowded schools, and other health-related issues. With these came the usual social problems: crime, prostitution, alcohol and drug abuse, violence, chronic poverty, and lack of education. To middle-class white observers, the fact that these areas were parsed ethnically made the correlation irresistible: blacks and foreigners, especially dark and Slavic foreigners, appeared prone to antisocial behavior and disease. On casual examination, heredity seemed coded in the social problems of the day; an average dose of garden-variety chauvinism was enough to confirm it.[26]

Science, technology, and medicine provided a new armamentarium against these social problems. The newspapers trumpeted a fantastic array of achievements and insights. Bell's telephone enabled one person to talk to another many miles away. Edison's electric globes turned night into day, and his phonograph preserved sound for later consumption, like salted meat or canned fruit. Marie Curie's radium seemed a source of unlimited and healthful energy. Roentgen's rays enabled him to see through solid objects and even into bodies. Darwin's bold theory of evolution challenged man's place in the cosmic order, suggesting that human beings and even societies were simply the result of natural processes. Koch, Pasteur, and Ehrlich had identified the germs responsible for man's great scourges; treatments and cures could hardly be far off. A huge public appetite for science stimulated a great deal of popular science writing, much of it straight from the front lines of research. From health to business to happiness, there seemed no problem insoluble by a rational, scientific approach. It was a golden age of scientism.[27]

Against this background, Mendel's laws seemed to have cracked the puzzle of heredity. Scientific enthusiasts quickly applied the new hereditary theory to such diseases as hemophilia, Huntington disease, and retinitis pigmentosa, as well as to hereditary predispositions to infectious disease. Galton's eugenic hypothesis—that with sufficient education, wise policy, and patience, we could guide our own evolution more gently and humanely than brute nature—acquired teeth.

The Progressives' infatuation with science implied an engineering approach to social problems: analyze a system into its components; identify those components' functions; repair or modify components as necessary. To understand a system was to have power over it. In the 1860s the great French physician Claude Bernard wrote, "When an experimenter succeeds in learning the necessary conditions" of the phenomena he is studying, "he is, in some sense, its master; he can predict its course and appearance, he can promote or prevent it at will." The turn-of-the-century biologist Jacques Loeb sought to develop experimental biology as a form of engineering; turning Bernard conversely, Loeb believed that promoting, preventing, and otherwise manipulating biological phenomena was a demonstration of true understanding. In 1914 the Baltimore physician Lewellys Barker introduced a collection of essays on eugenics, writing, "Wherever man has begun to know scientifically, he has found himself also, better than before, able to predict; he has gained the power to control." The desire for hereditary engineering lies in the roots of biomedicine.[28]

These biologists' wish for control dovetailed with many Progressive initiatives, but it was particularly conducive to emerging ideas about health. Among these reform movements, health was one of the key themes. Health reform encompassed campaigns to improve public facilities, products, and services such as sanitation, food, drugs, and working conditions; activist groups combating specific diseases, such as tuberculosis or syphilis; and many "personal reform" movements that promoted particular diets or ways of living. Some health reforms, such as Country Life, Fletcherism, and Clean Living, were cultish and alternative; others, such as the lobbying effort to establish a national department of health, were mainstream and bureaucratic. Some groups, such as the temperance and antisaloon leagues, were older and socially conservative, while others, such as the birth control movement, were newer and more liberal. Many of these reforms went under the banner of "hygiene." It has been called a time of "strange theoretical combinations," when many odd bedfellows were made. And thus eugenics advocates found themselves allied with health reformers of many stripes, promoting a hygiene of the germ plasm.[29]

* * *

By 1906 Irving Fisher was a vigorous participant in Progressive health reform. He was an antituberculosis activist and an ardent prohibitionist, with an interest in heredity inculcated by Kellogg and nurtured by his own reading and research. That year he heard a valedictory paper by J. Pease Norton, a fellow Yale economist, the outgoing vice president of the Economics Section of the American Association for the Advancement of Science. Norton's paper both epitomized the Progressive spirit—the sense of duty to advance civilization by means of expertise, organization, and efficiency—and evoked the now-familiar eugenic language of coercion and "race suicide":

> The salvation of the civilization and the race lies in the hands of exceptional men. The hope of the race inheres in their efficient organization for action. Organization consists in compelling voluntarily or involuntarily each individual to do that thing within his capability which has greatest value for society. To do otherwise is a great waste. To permit great wastes to go unchecked is more than a suicidal policy; for an evil more heinous than race suicide is race homicide.[30]

The great wastes Norton identified were preventable death, preventable sickness, preventable conditions of low physical and mental efficiency, and preventable ignorance. Disease, deformity, insanity, feeblemindedness, and lack of education. Progressive-era health reformers perceived eugenics as addressing the hereditary side of these near-universal concerns, but each of these was understood, then as now, as having roots in both heredity and environment. The logical Progressive solution was bureaucracy: Norton called for the establishment of a national department of health. A little later, the movement here begun would pick up the widespread agricultural metaphor, arguing that a national health department would "spread throughout the country a knowledge of effective ways of stamping out disease, as the Department of Agriculture has done with cattle."[31]

By August, Norton's exhortation had become Fisher's battle cry. Fisher began to organize a committee to press for health reform and especially to lobby for the creation of a national, cabinet-level Department of Health. In doing so, he called upon and knit together a social network of health reformers. This ensured a role for medical eugenics within the structures of Progressive health reform. He began locally: his first recruits were those present at the AAAS session, Yale colleagues and fellow alumni, associates and health reformers, and fellow board members from organizations such as the American Breeders' Association. Nearly every member of the committee was a friend or a friend of a friend of Fisher. Whimsically acknowledging the inbred nature of the fledgling group, Fisher wrote to his longtime friend the Rev. William G. Eliot, Jr., "I 'deny the allegation of nepotism in getting up this committee and defy the alligator!' "[32]

Fisher then worked outward, calling on friends, colleagues, and personal acquaintances up and down the East Coast, seeking people from a range of fields with the status, power, and access to influence the national debate on health. On the West Coast, he knew the agronomist Luther Burbank and Stanford University president David Starr Jordan from the ABA committee, and Eliot, who lived in Oregon, helped swell the list with westerners. "I don't want it much larger," Fisher wrote, but the list continued to grow. He added more physicians, including the antituberculosis crusader S. Adolphus Knopf, and brought in Trudeau, Kellogg, Horace Fletcher (the Great Masticator himself), and more biologists, such as the staunch mechanist Jacques Loeb. Wisely Fisher also added philanthropists such as Henry Phipps, who founded clinics at Johns Hopkins and the University of Pennsylvania, and administrators such as John Shaw Billings, a surgeon and the founder of the New York Public Library; Hermann Biggs, the New York State Commissioner of Health; the psychologist G. Stanley Hall, president of Clark University; and Charles Eliot, the former president of Harvard. He also attracted such celebrities and public figures as Thomas Edison and Booker T. Washington. In short, Fisher gathered much of the American intellectual elite, scientists, physicians, health reformers, and money makers. The roster was complete by the end of the year, and the group soon had stationery, officers, and offices in New York and New Haven. It became known as the Committee of 100.[33]

Where Norton's original address had stressed an agricultural theme, the Committee of 100 adroitly shifted the metaphor from husbandry to the environment. President Theodore Roosevelt charged the group with extending the notion of conservation of environmental resources to that of conservation of human resources. (The conservation movement had its own links to eugenics, especially in California, with groups such as the Save the Redwoods League.) Although Roosevelt publicly supported the Committee, he wrote Fisher privately that he in no way supported the Committee's principal aim of adding to the federal bureaucracy. "I emphatically disapprove of a Cabinet officer being created at the head of a Department of Health," he told Fisher, "so please do not use my letter at all if your body concludes to agitate for a Department of Health." TR was willing to consider a Bureau of Health, under the Department of Interior or perhaps Agriculture, "but we need no additional Cabinet officers." The choice came down to metaphors. Putting a bureau of health under the Agriculture Department would exploit the comparison with breeding, while putting it under Interior would draw

2.5 Irving Fisher, Thomas Hunt Morgan, Alexander Graham Bell, 1915. Courtesy of Cold Spring Harbor Laboratory

on the metaphor with natural resources. Health was a resource to be conserved, enhanced, managed, and wisely exploited, just as timber, minerals, or fresh water. Physicians—and eugenicists—were the foresters of the human timber, charged with using the latest in scientific knowledge to intelligently and benevolently manage human vitality and efficiency for the maximum national benefit.[34]

Under the Committee's auspices, in 1909 Fisher wrote the *Report on the National Vitality*, a 138-page essay that marshaled his skills as an economist. If we appraise a human life "at only $1700," he wrote, "and each year's average earnings for adults at only $700, the economic gain to be obtained from preventing preventable disease, measured in dollars, exceeds one and a half billions." A staggering amount to Progressive ears. After delineating, with a welter of statistics, estimates of human life expectancy and the causes of mortality, Fisher adumbrated the means by which this prodigious savings could be had. First on the list was "Conservation through heredity." "Human vitality depends on two primary conditions," Fisher wrote, "heredity and hygiene. . . . In other words, vitality is partly inherited and partly acquired." He invoked the agricultural metaphor almost reflexively, noting that the original apple, "as offered by nature to mankind, was the small, sour, bitter crab of the forest, unpleasant, indigestible, innutritious." "Human heredity," he continued, "is now based on haphazard selection." He followed with a survey of eugenics from Galton to Pearson, but also observed that Mendel's theory needed to be studied for practical application. He noted that Galton's original model of eugenics called for nothing more than education of the masses as to the universal benefits of eugenic mating, but he was sanguine about eugenic laws such as had recently been passed in Indiana, Connecticut, and Michigan. Fisher devoted chapters to conservation through public hygiene (clean water and vaccination to promote public health, for example), semipublic hygiene (in schools, businesses, and the like), and personal hygiene, indicating that these ranked with eugenics as fundamental levels of "vitality." He then returned to eugenics to address the obvious question of whether it conflicted with hygiene. "It is charged that hygiene prolongs the lives of unfit and defective classes," he acknowledged. On this point Fisher sided with Francis Galton—and departed from the social Darwinists, including his mentor Sumner, who used this logic to argue against any sort of social welfare programs or education. While admitting that humanitarian impulses that betrayed us into "shortsighted kindness" must be checked, nevertheless he maintained that "all the dangers of perpetuating vital weaknesses can be avoided if proper health ideals are maintained." Eugenics and hygiene, he believed, must work together to reduce the numbers of children born to degenerates and defectives and to raise the public consciousness regarding the value, both economic and aesthetic, of health, vitality, and efficiency.[35]

The Committee's program took shape in the form of a bill introduced to the Senate by Senator Robert L. Owen. Fisher took courage from the recent election of his old friend William Howard Taft to the presidency. But the Owen bill had the bad fortune of being introduced in a year when at least six major health reform bills competed for the attention of Congress. The Owen bill lost to a competing bill supported by Surgeon General Walter Wyman. The Wyman bill reorganized the Public Health and Marine Hospital Service into the Public Health Service and significantly raised the PHS budget, although it remained under the Treasury Department.[36]

After that defeat in 1912, the Committee of 100 broadened its agenda to tackle other health-related causes, while still, of course, pushing for a federal bureau or agency. They supported the Harrison Cocaine and Opium bill of 1914, which would have strengthened interstate regulation of narcotics by requiring dealers to have a license, and they favored a bill that would sequester lepers in institutions. "In some States they are allowed to wander unrestrained," wrote one advocate. "In most States the unfortunate town which discovers them must assume the burden of their care, although they are generally foreigners." Eugenic thinking was not restricted to purely eugenic problems. Fisher persevered with the Wilson administration through the teens, but to no avail, due largely to Wilson's increasing distraction with the war. The Owen bill was revived and reintroduced in 1919 but died in Congress. The Committee of 100 continued on paper, but its fundraising and activities dropped to zero.[37]

The Committee of 100 failed in its stated mission, but it had a powerful influence in coalescing a broad social network of health reformers, men and women with overlapping interests, each subscribing to a constellation of social causes. Besides his prohibition and anti-tuberculosis work, Fisher donated to the Tuskegee Institute, the Birth Control League, and the American School Hygiene Association, and he later lobbied on behalf of the League of Nations. Eugenics leaders were on the boards of charities for other causes, such as public health, education, or temperance, while activists for those causes often supported eugenics. The birth control activist Margaret Sanger was a well-known eugenics supporter. In her 1922 treatise *The Pivot of Civilization*, she wrote, "There is but one practical and feasible program in handling the great problem of the feeble-minded. That is, as the best authorities are agreed, to prevent the birth of those who would transmit imbecility to their descendants." Adolphus Knopf founded what became the

American Tuberculosis Association, and he actively supported eugenic policy and legislation and served on the boards of a number of eugenic committees and associations. Similarly, his board members Edward L. Trudeau and Hermann Biggs joined eugenics organizations. Billings served on the boards of several eugenics societies as well as the American Public Health Association. The Johns Hopkins psychiatrist Adolf Meyer was on the board of the National Committee on Mental Hygiene (founded by Fisher) and found time to serve on the boards of several eugenics societies, the Bureau of Social Hygiene, and the American Birth Control League. Heredity and hygiene were overlapping domains. For some they were the poles of a spectrum; for others, one was a subset of the other. In the world of Progressive era health reform, eugenics was recognized as one tool in the kit; everyone recognized heredity as playing some role in nearly every aspect of health, and eugenics claimed to address that piece of the puzzle.[38]

Even the American Breeders' Association was part of the health reform network. The Eugenics Committee's charter roster included Welch and Meyer, then in charge of the Manhattan State Hospital on Ward's Island, in New York City's East River; Welch would soon recruit him to Hopkins. The Eugenics Committee grew and was converted into a section. By 1912, it encompassed committees on the heredity of insanity, epilepsy, deaf-mutism, eye defects, mental traits, and feeblemindedness, in addition to those on sterilization, criminality, genealogy, and immigration. Many of these committees were headed by M.D.s. From the moment of its inception, then, professional American eugenics was applied human genetics, a science that encompassed a wide range of interests, prominent among which were those concerned with human health. From the ABA's Willet M. Hays, Meyer requested permission to forward ABA materials to Alfred Plötz and Ernst Rüdin, German geneticists involved in the International Association of Race Hygiene, which Meyer had also joined. In the 1930s, Plötz and Rüdin would become leaders of the Nazi eugenics program. But of course Progressives had no foreknowledge of eugenic abuses. Eugenics harbored a core of enthusiasts, but other reformers were equally chauvinistic about alcohol, poverty, tuberculosis, or unclean food as the root of all evil. It was a zealous age. Common to all was an effort to take a rational, scientific approach, not to let sentiment cloud one's judgment, to do the most collective good possible.[39]

Indeed, to a first approximation, everyone in the Progressive era was a eugenicist. Eugenics tapped widespread Progressive concerns with purity and improvement. Reform-minded Progressives saw eugenics as addressing the hereditary dimension of the same sorts of social problems all reformers were interested in: morality, safety, education, immigration, race, and disease. Essentially all reformers accepted eugenics as one strategy toward reform, though only a few thought it the only or even the most important strategy. Those most strongly hereditarian in outlook naturally included those who studied heredity professionally. And health, being universally acknowledged as having a hereditary dimension, was highly conducive to eugenic arguments. Nearly everyone believed disease and antisocial behavior were at least partly hereditary; one's commitment to eugenics indicated the extent to which one believed that social problems were innate. Eugenicists who believed that health (like everything else) was mainly hereditary avidly attached themselves to health reform movements in order to advance their agenda; at the same time, health reformers embraced eugenics as one of multiple means to longer, healthier, happier lives. The belief that everything human was hereditary was simply an extreme position on a spectrum on which every reformer found her own comfort zone.

Many of these health reform movements gathered under the widening umbrella of "hygiene." In the Progressive era, the nineteenth-century idea of hygiene—bathing, grooming, avoidance of spitting, and so forth—came often to be referred to as "personal hygiene." The germ theory of disease gave scientific muscle to the old sanitarian rhetoric about cleanliness, and religious imagery continued to be potent: a "gospel of germs" guided behavior from comportment to home maintenance to the purchase of appliances and bathroom fixtures. As hygiene spread it multiplied and developed nomenclature. Under the rubric of "semipublic" hygiene were lumped school hygiene, work hygiene, industrial hygiene, and location-specific hygiene. More broadly still, "public" hygiene connoted cleanliness of city parks and sidewalks, train stations and stores. In 1909 Adolph Meyer suggested the term *mental hygiene* to denote the care of one's faculties and especially the preservation of sanity. *Race hygiene* may today evoke chilling Nazi associations, but the German *Rassenhygiene* long predates National

Socialism. In Progressive-era America race hygiene connoted a wide range of social behaviors intended to make the germ plasm wholesome, healthy, and clean, from public health to "intelligent marriage." Illustrating the linkage between mental hygiene and race hygiene, Meyer gladly accepted an invitation to join the English Eugenics Education Society in 1909, stating confidently, "There will, I am sure, be many persons interested [in the EES] in this country and esp. those of us who lately founded the Nat. Comm. for Mental Hygiene." Although some aggressive eugenics advocates claimed that health reform weakened the genetic stock by promoting the survival of the feeble, some leading voices—including Fisher's—argued that hygiene and eugenics were components of the same larger project.[40]

Medical eugenics, like all hygiene, was preventive medicine. In a time characterized by great faith in rationality and yet relative therapeutic impotence, scientific approaches to hygiene, nutrition, regimen, and heredity were frontline defenses against disease. "Modern medicine, yielding to the demands of real progress, is becoming less a curative and more a preventive science," wrote Harvey Ernest Jordan, Dean of the Medical School at the University of Virginia, in a talk delivered in June 1912 and published later that year. Infused with the Progressive spirit, he continued: this shift toward prevention "represents the medical aspect of the general change from individualism to collectivism." The rational, Progressive approach to diseases was to strive for "their racial eradication rather than their personal palliation." Jordan believed that the "possibilities now exist for developing a physically, mentally, and morally stronger and healthier race," and that the existence of those possibilities demanded action. It was, he said, the compassionate thing to do. "Enlightened society demands the elimination of as much of the physical, mental and moral sickness and weakness as can be prevented." The bureaucratic, scientistic approach so characteristic of the age put sleek new clothing on the old spirit of human perfection: "The ultimate ideal sought," he wrote, "is a perfect society constituted of perfect individuals." Jordan's vision of eugenics imported the nineteenth-century vision of holistic purity into a modern, technological, biomedical, Progressive worldview.[41]

In the same month, Jordan's friend and colleague Charles Davenport articulated the way eugenic medicine spanned all social levels, from the individual to the population. Addressing the Brooklyn Medical Society, he wrote that doctors have "peculiar and peculiarly intimate relations, both with individuals, with families, and with communities." Treatment, of course, begins with the individual. The physician knows that each patient is an individual and requires unique attention. Accurate treatment must always take into account the patient's family history. Also, knowledge of the family can help the physician distinguish hereditary from acquired disease and can inform diagnosis and therapeutics. Finally, the physician has a vital role in the community, as an adviser on work, nutrition, marriage, and child-bearing. Davenport believed that the physician's intimate relation to the state, that is, to organized society, gives him a special responsibility. Defects—especially mental defects—"often cause great loss of property and life." Building to an evangelical crescendo, Davenport wrote, "If we persist in our crazy policy of protecting the weakest strains from the action of a natural selection, while permitting them to breed, our nation, too, will soon be numbered among those that were great." Thus, for Davenport, there was no conflict between the Garrodian concern for the individual and the Galtonian concern for the population.[42]

* * *

Fisher's next health reform project united the social network of the Committee of 100 with his long-standing ties to John Harvey Kellogg and the San. In the closing years of the nineteenth century, Kellogg had created a charity called the American Medical Missionary and Benevolent Association—a nod to his Adventist roots and to the enduring relations between piety and health. The AMMBA eventually became the Race Betterment Foundation, which had a more modern, scientific ring that harmonized better with "biologic living" and was more congenial to staunch eugenicists. The most important activities of the Foundation were three conferences, in Battle Creek in 1914, in San Francisco the next year, in conjunction with the Panama Pacific International Exposition, and again in Battle Creek in 1928. A fourth conference was planned for 1940, but was scuttled because of the Second World War.[43]

The First Annual Conference on Race Betterment was a quintessentially Progressive project that illustrates the integration of medical eugenics into the wider health reform movement. Fisher was a key figure. He was vice president of the conference, a member of the executive board, and a contributor to the proceedings. The roster of participants—some four hundred—was impressive, both in distinction and diversity. The organizers attracted celebrities such as Booker T. Washington; Jacob Riis, the muckraking author of *How the Other Half Lives*; and Gifford Pinchot, the conservationist and first chief of the U.S. Forest Service, as well as J. McKeen Cattell, the statistician and science publisher. Eugenicists were on hand, of course—men such as Roswell Johnson, a biologist from the University of Pittsburgh and a close associate of the Eugenics Record Office, and Paul Popenoe, then serving as editor of the *Journal of Heredity*, at the time the primary American journal for genetics and eugenics research. Four years later, Johnson and Popenoe published *Applied Eugenics*, which became the most widely used college textbook on eugenics. More than thirty physicians attended—even discounting Davenport, who for at least the second time in three years represented himself as an M.D. The antituberculosis crusader S. Adolphus Knopf participated actively, as did the judge Victor C. Vaughan, who was also president-elect of the American Medical Association. Vaughan lived in nearby Ann Arbor, where he headed the state medical society, and he was an avid supporter of eugenics, cozy with delegates such as Davenport and Laughlin of Cold Spring Harbor, and, of course, Kellogg. In addition, the conference welcomed more than a dozen social workers, a number of people involved in publishing, politicians from the local, state, and federal levels, and members of the clergy. The roster of delegates was, in short, a who's-who of Progressive health and purity.

To most of the delegates, "race betterment" was neither "racial," as we use the word today, nor exclusively eugenic. *Race* usually denoted the human race, and betterment was any means, hereditary or environmental, by which degeneration could be arrested and reversed. Calls for national birth and marriage registries, for programs of health certification, and for more and tougher marriage and immigration laws reflected the Progressive faith in bureaucracy. There was much talk of prohibition, both of alcohol and of tobacco. Sex, of course, was on many people's minds: both how to keep children from having it and how to keep adults from having it too often, with too many, or the wrong, people. Aside from a few diatribes, the overwhelming tenor was one of rationality, of logical solutions to social problems. Rational people ought to avoid deliberately ingesting poisons such as alcohol and nicotine. Rational people ought to subjugate passion for intelligent marriages that will improve future generations. And rational people ought to cultivate compassion for the disadvantaged, to help them realize their full potential to contribute to society. The tone of the proceedings is sanctimonious, repressive, and, with a few exceptions, compassionate.

Universally, the delegates believed that the human race—in a few cases, the "American race"—was falling apart. The sources of degeneration were endemic in modern society: pollution, rapacious business, corrupt government, immigration, immorality, temptation, and laxity. The conference's stated aim was to describe those forces of degeneration and to begin amassing suggestions to counteract them. In almost everyone's mind, heredity was part of both the problem and the solution, but only a handful of enthusiasts believed it was anything like the entire story. Degeneration was both physical and mental. "It is within the power of man to modify his environment," wrote Kellogg. "Destructive, degenerative tendencies" could be eliminated through discipline, regimen, and rationality. In time, this would produce a "race of thoroughbreds" as superior to the current average man as the equine thoroughbred was to the average horse.[44]

The participants sought to stem and reverse this degradation by scrubbing their skins, detoxifying their bodies, decontaminating their environments, and purifying their germ plasm. Hygiene at all levels was discussed at length: clean bodies, clean homes, clean workplaces and schools, clean food, clean mind. Moral cleanliness continued to be just as important as physical cleanliness. A tightly buttoned morality pervades the proceedings. The physician Daniel Lichty called tobacco a "race poison." Temperance advocates spoke out against the "rising tide of alcohol consumption," the "sacrifice of boys and girls" to alcohol, and the "booze special," a euphemistic train to Hell. Sex, of course, remained titillating and appalling. One can almost hear the quiver in the voice of J. N. Hurty, state health commissioner for Indiana and an architect of the state's sterilization law, warning of the "very strong, exceedingly strong, sexual appetite of men." Winfield Scott Hall advocated sex education in schools (for adolescents), but saw it as moral and eugenic training rather than as a pragmatic step to reduce

2.6 Banquet of the First International Conference on Race Betterment, Battle Creek Sanitarium. Courtesy of American Philosophical Society Library

pregnancy and sexually transmitted disease. Indeed, if all men were virtuous, he wrote, we would eliminate venereal disease—a source of degeneration doubly potent because it stood for both physical and moral decline. Race betterment meant clean living, country life, vigorous exercise, abstinence from recreational toxins, and tightly regulated sex. Indeed, the conference proceedings consistently imply that no place fosters race betterment as effectively as the Sanitarium at Battle Creek.[45]

Pervading the conference, then, was an ethos of perfection, inside and out. Hurty wrote of the "elements of ideal, perfect manhood" and womanhood, "perfect physically, mentally, and morally." Theirs was an Edenic conception of human nature, in which a pure ideal form could only be eroded by modern life. The climax of the conference was the awarding of medals for two "physical and mental perfection contests" carried out over the previous months on the children of Battle Creek. All children were measured, weighed, examined, and tested for every physical and mental quality. Gold medals were awarded for the best in each category. "Perfection" sometimes meant "best" and sometimes "most normal"—in other words, sometimes for the right-hand tail of Galton's bell curve and sometimes for the peak in the middle. Teeth should be as straight and blemish- and cavity-free as possible. Perfect vision, too, implied maximum, not average, sharpness. But height-to-weight ratios ought to be normal, not too extreme. Common to all tests is a notion of eliminating deviation—echoing elsewhere the abhorrence of *deviance*—whether that deviation is a degradation or an extremism. Along with the high-mindedness went a profound sense of conformity, the longing to be a cog in an oiled machine running inexorably toward perfection. The contest epitomized Jordan's shift from individualism to collectivism.

The delegates struggled to define the relationship between nature and nurture. For Irving Fisher, nurture equaled hygiene and nature equaled eugenics. "There are two factors which cooperate to produce vitality," he wrote, using his preferred term for that quality whose increase led to betterment, "namely, heredity and hygiene." He sought to put to rest the idea that the two were somehow at odds. One needed both. Hygiene, he wrote, was "merely a common handicap for all classes." His program for race betterment included everything from eugenic marriage laws and sterilization, to elimination of immorality, to higher ideals of health and vitality. Heredity was critically important to Fisher, but it blurred insensibly into environment, nurture, and hygiene. Statements on nature and nurture recur throughout the text; the phrase was on everyone's tongue, and presenters seem to have vied for the pithiest summaries and most apt characterizations of the dynamic between heredity and environment. "When I am asked, as often happens," wrote Leon Cole, "which I consider of greater importance, heredity or environment, I commonly give a Yankee reply by asking in return, Which is of more importance for sustaining life—food or air?" Even if such statements were mere lip service, one was apparently expected to do at least that much.[46]

Heredity and environment were more than complementary; they could become deeply entangled. Several of the speakers, including Kellogg himself, discussed the inheritance of acquired characteristics. In the Progressive era, heredity could include "environmental" influences we would today call congenital; conversely, environmental influences were widely believed to shape heredity. At Battle Creek, the antitobacco crusader

Daniel Lichty evoked an image of sickly, yellowed germ plasm: "Nicotine begets very decidedly neuropathic stock. The heredity of nicotine-tainted stock is never on the right side." Conversely, social actions commonly associated with eugenics could be advocated for on purely "environmental" grounds. S. A. Knopf considered it "well-nigh a criminal offense" for tuberculous people to have children, even though he understood perfectly well that hereditary transmission of tuberculosis is rare at best. Extreme hereditarians were few at the conference, and most delegates acknowledged them as loose cannons who could be more destructive to the cause than helpful. A softer, more permeable conception of heredity suited health reform better.[47]

Contemporary concepts of heredity played perfectly into the notion of purity. Hurty sang of "blemishless heredity," implying an ideal set of traits that could be sullied only by environmental insult or unwise mate choice. Such a concept is in fact implied by the genetics of William Bateson. *Principles of Mendelian Heredity*, the masterwork of Thomas Hunt Morgan and his *Drosophila* boys, would not appear until the following year. Bateson's "unit character" still held sway, and few undogmatic experimental geneticists attended the Battle Creek symposium. The unit character, and its sister concept, the presence-and-absence theory, implied in cellular terms the notion of virginal purity exploited by the Battle Creek delegates. Bateson's concept of "allelomorphs," or alternative forms of a gene, was one of normal and abnormal. The "germ plasm," the Progressive-era term for one's total heredity, was conceived as an ineffable substance whose ideal state was perfect purity. It could only be contaminated, in a hundred different ways. It was a restrained, utilitarian moment in the science of human perfection, lacking the utopian playfulness of either Galton three decades earlier, or the twenty-first-century eugenic visionaries a century hence.

* * *

The second Conference on Race Betterment, held the next year in San Francisco, noticeably deemphasized euthenics, or improvement of the environment, compared with eugenics. It thus signaled, writes the historian Alexandra Minna Stern, a falling tide for holistic race betterment advocates such as Kellogg. The cereal king's enthusiasm for vegetarianism and hydrotherapy isolated him from the eugenics ruling class and has largely kept his brand of human improvement out of the main narratives of eugenics and human genetics. Part of Fisher's significance is that he carried the torch for Kellogg, on into the 1920s, importing a version of health-centered hygienic holism into the center of the eugenics movement.[48]

In the late teens, Fisher's health reform efforts merged more closely with his work as an economist, and the two took a practical turn in the creation of a company. He recognized the notion of health conservation as being of intense interest to insurance companies. In 1913 he and his Yale colleague Eugene Lyman Fisk, M.D., created the Life Extension Institute, "a self-supporting central institution of national scope devoted to the science of disease prevention" and to "preach[ing] the gospel of individual health." The LEI supported itself by providing low-cost medical exams to the public and selling the results to insurance companies. Its staff did not treat patients, however; if a medical condition was found, the patient was advised to see his or her doctor. Fisher drew heavily from his network and their connections, in terms of principles, agenda, and membership. The Institute's Hygiene Reference Board, its primary board of advisers, comprised a large roster of prominent physicians (among them Welch, Lewellys Barker, George Crile, and William J. Mayo), scientists (Walter B. Cannon, Lafayette Mendel), health reformers and public health researchers (J. H. Kellogg, Theobald Smith, Charles Stiles), eugenicists (Davenport, Bell, David Starr Jordan), and philanthropists (the Rockefeller's Wickliffe Rose). Fisher persuaded William Howard Taft, who had some time on his hands as a result of having recently been trounced in his presidential reelection bid, to serve as chairman of the board. Taft, a lawyer, also advised on legal matters, which the LEI needed; it regularly ran afoul of state laws prohibiting the practice of medicine by private corporations. Accusations of improper financial relationships caused considerable distress in 1920. When Welch resigned from the Hygiene Reference Board, citing objections to their practices, Fisk protested like an unwisely honest child, denying involvement in recent scandals that Welch had not mentioned: "I do not know whether the criticisms that have come to your ears relate to alleged fee splitting on the part of our x-ray laboratory, which is wholly erroneous, or to our alleged connection with the anti-tobacco propaganda." Some scandals had apparently gone public. "An article recently appeared in the Atlantic Monthly which was so grossly unfair to the Institute that I demanded of Mr. Sedgwick some public retraction. It is quite obvious to me that this article was written by some one in the employ of the tobacco interests." The LEI repeatedly denied that it was practicing medicine, drawing a fine line between diagnosis and therapy.[49]

Fisher used the LEI as a vehicle for championing hygiene in all its guises. Under LEI auspices, Fisher published, with Fisk, a self-help manual of hygiene. Paternalistically titled *How to Live,* it conveyed a sense of medical authority to Progressive-era readers. The book is Fisher's 1909 *Report on National Vitality* expanded, diluted, and repackaged as self-help. It emphasizes cleanliness, lots of roughage in the diet, ample mastication, and sleeping outdoors. Though it is primarily a manual of "individual hygiene," its concluding chapter emphasizes the importance of semipublic, public, and, "the most important of all," race hygiene.[50]

To call race hygiene the most important form of hygiene was to risk invalidating all the advice on healthy living that had come in the preceding pages. The tension between nature *versus* nurture and nature *and* nurture never goes away: "In a general way, there is a broad distinction between eugenics, which is the hygiene of future generations," the authors noted, and public and semipublic hygiene, "which relate to the present generation." "It is true that if followed out faithfully, the rules of hygiene will enable a man to live out his maximum natural life-span, with the maximum of well-being, and to run no risk of allowing any inherent weakness to be brought out." However, they said, a good environment will take one only so far. "Some persons, even if they followed what is very nearly the normal code for the human being, would scarcely be able to avoid dire physical and mental fates. In short, we find that besides the hygienic factor in life which we may call environment, there is something else on which the health of the individual depends. This something else is heredity, or 'the nature of the breed.'"[51]

Thus, by rejecting a strict opposition of nature and nurture, Fisher and Fisk hybridized public health with animal husbandry: eugenics is that branch of hygiene concerned with the nature of the breed, with all that connotes. Fisher and Fisk advocated standard eugenic solutions to poor race hygiene: prevention of reproduction by the "markedly unfit"; enactment of "wise" marriage laws; and "development of an enlightened sentiment against improper marriages," that is, education (or propaganda). Nature and nurture must work together, but even in this self-help manual for clean living, nature trumps. "If we are to build for future generations," they wrote, "hygiene must give way to, or grow into, eugenics." Genetic improvement, in short, was the ultimate in preventive medicine.[52]

In concert with his views, Fisher's health work placed increasing emphasis on eugenics. In 1921 he attended the second International Eugenics Congress, organized by Davenport and Henry Fairfield Osborn and held at New York's American Museum of Natural History. The museum was the home court of the Galton Society, and the meeting's title, Eugenics in Race and State, reflects their ruling-class style of eugenics. Alexander Graham Bell, one of the movement's greatest public relations assets, was the honorary president. It is in many ways the climactic document of the eugenics movement proper (capital-E eugenics, as one geneticist would later say). And Fisher, by this time, was firmly in the inner circle. The movement's epicenter was located somewhere in the middle of Long Island Sound, roughly equidistant from Cold Spring Harbor, the Upper West Side, and New Haven.

Despite the overall anthropological tone, this meeting, like all eugenics meetings, contained references to health and medicine and calls for more physicians to support the cause. Stewart Paton of Princeton University echoed the sentiments of Davenport and Jordan a decade earlier. The physician, he wrote, did not yet grasp his crucial role in improving the quality of human thought. Though he acknowledged how little was as yet known about the principles of heredity, Paton remained optimistic. The time for a true positive eugenics was coming soon. "Recently we have been thoroughly impressed with the importance of eliminating the unfit," he wrote, "but how much attention has been given to the assistance of the *fit* toward the full stature of their development?" Physicians, he wrote, were in a unique position to assist with the development of human beings toward their full potential—toward, in short, perfection. Following a veritable cavalcade of waffle words—"We do not believe it an idle dream that the time is not very distant when . . ." —he got to his point: medicine was crucial to the Progressive project of scientistic social engineering. "Our law makers and statesman [*sic*]," he continued, "will realize the necessity of learning something about the nature of the forces controlling human beings. Then clinics for the study of human behavior will be frequented not only by medical students and those interested in the problems of eugenics and social welfare, but by every person who has an intelligent interest in learning how to control and to direct the energy of human endeavors."[53]

Within two decades, such clinics did open. Beginning in 1941 they were indeed frequented by medical students and those interested in eugenics, and they did evolve into institutes that sought to explore and manipulate the forces controlling human beings (chapter 4). And eight decades later, physicians would begin to reimagine their role as helping all citizens, not just the sick, attain their full potential.

At the climax of the 1921 Eugenics Congress, the delegates formed a committee to promote eugenics education. The committee roster illustrates the linkage among the health movement, eugenics, and academic genetics: Fisher (chairman); Davenport (vice chairman); Clarence Cook Little, a human and mouse geneticist who went on to found the Roscoe B. Jackson Memorial Laboratory in Bar Harbor, Maine (later Jackson Laboratory); Madison Grant; and Harry Olson, a municipal judge in Chicago and an architect of eugenic sterilization laws. This committee became, first, the Eugenics Committee of the United States of America, and then, in 1925, the American Eugenics Society. Fisher was its president from 1922 to 1926, when it was incorporated.[54]

In 1923 Fisher made clear the linkage between the new Eugenics Society and his earlier public health work. In what was probably the last official memorandum of the Committee of 100, he wrote that one of the goals of the fledgling eugenics society was to establish a federal Department of Health. The Committee of 100 disbanded, donating its small treasury to the Eugenics Committee. As president of the AES, Fisher worked closely with Davenport. The American Eugenics Society and the Eugenics Research Association were complementary sister institutions. The ERA's mission was to support eugenics research, while that of the AES to support eugenics education and propaganda. Together, they formed the backbone of American eugenics in the 1920s.[55]

* * *

Fisher's quest reveals the connections among spiritual and physical purity, public health, Progressivism, genetics, and eugenics. Like other health reformers, Fisher was deeply interested in heredity and eugenics yet unconcerned with Mendelian genetics. Identifying dominant and recessive unit characters mattered less to him than toning the human constitution by whatever means available. This soft sense of heredity, then, is a key theme in the early history of medical genetics. It is here, where eugenics is a form of hygiene, where heredity and environment are complementary, and where prevention and predisposition guide medical practice, that genetics slipped into medicine. Notions of purity and perfection conditioned the minds of those interested in heredity and health, imbuing them with complex and often contradictory attitudes toward human betterment. A pure, ideal germ plasm would be disease-free, ever healthy—ever young.

This public health–oriented eugenics has tended to be buried beneath the more sensational and horrific history of eugenics, much of which had nothing to do with health or medicine and everything to do with a racist, elitist ideology. To nuance that history with this gentler narrative of public health eugenics is not to forgive bad science or cruelty perpetrated in the name of genetics. Rather, it makes the strange disconcertingly familiar. Shifting our gaze from the dogmatic, pseudoscientific propaganda of the worst eugenic abuses to the legitimate desire to use our knowledge of heredity in the fight against disease denies the easy righteousness that precludes reflection. It forces us to confront the eugenic impulse that runs through the recent history of heredity and health. That impulse manifests in manifold ways. Its expression depends on cultural context, social goals, technical knowledge, and personal philosophy. It is not in itself evil, though it is often insidious. Tracing the eugenic impulse through the changing times has at least two implications for how we view contemporary biomedicine. It relieves us of the fear that Progressive-style eugenics will return, because that would require repetition of the Progressive era itself—a phenomenon far too complex to happen twice. But simultaneously it makes the familiar strange, by challenging us to examine uncomfortable questions about individuality, autonomy, determinism, and social control in medical advances presented as benign. That cannot return which never went away.[56]

3

A Germ Theory of Genes

"DEAR SIR," THE LETTER BEGAN, "I have been studying migraine and its inheritance in the population of Western North Carolina." The year was about 1927, and the recipient, Laurence Snyder, was a young geneticist still finishing his Ph.D. at Harvard but already on the faculty of North Carolina State College in Raleigh. The letter's author was William Allan, a country doctor 170 miles away in Charlotte. "My results seem to indicate that migraine is due to a dominant factor," Allan continued. Snyder may have cocked an eyebrow at the old-fashioned term *factor;* professional geneticists had long since adopted Wilhelm Johanssen's term *gene*. Snyder's correspondent was no sophisticated geneticist. As Allan lays out his results, his expectation of a simple Mendelian dominant seems increasingly far-fetched. The trait occasionally appears in children even when neither parent is afflicted. When both parents have the trait, Allan expects a 3:1 ratio (true only if both parents are heterozygous), yet he gets closer to 5:1. He uses the term *backcross* inappropriately and anyway does not obtain the expected ratio there, either. Snyder must have smiled when Allan concluded, "Does this mean that I do not have the right to consider migraine as a simple dominant?"[1]

Yet the country doctor was on to something. With Snyder's guidance through some sticky statistical manipulations, Allan teased out from the messy data what did indeed appear to be a dominant allele that contributed to migraine. The results looked nothing like Gregor Mendel's uncannily perfect ratios, but human pedigree data usually require a good deal of mathematical filtering to separate the signal from the noise. Allan the physician and Snyder the geneticist were stumbling toward a science of medical genetics.

Hesitantly, with false starts and occasionally misguided efforts, academic genetics integrated with medicine. It traversed the chasm separating the two disciplines by means of a few pioneers who made the trip themselves or, like Allan to Snyder, lobbed over a weighted line in the form of a letter. Their motivations were many, but two main approaches emerged. Some thought of heredity as constitutional, part of the soil. They sought, by means of statistics and various theoretical contrivances, to address health and disease holistically. Others treated genes as the seeds of disease. They played off the perceived successes of the germ theory of disease to articulate what we might call a germ theory of genes—a medical genetics based on the idea of genes as agents of disease. These two approaches competed through the nineteen twenties and thirties, and for a time it was not obvious which would predominate. Both seed and soil approaches helped push the preventive, public health message of Progressive medical eugenics into mainstream academic medicine. But the germ theory of genes was better adapted to its cultural moment: it was easier to understand, easier to link to past medical successes, and easier to sell to funders and to the public.

* * *

In the late 1920s, applied human genetics was still synonymous with eugenics, and human genetics was still practiced by Ph.D. geneticists in arts or agricultural schools, such as Harvard's Bussey Institution, the Agricultural Experiment Stations in Maine, Connecticut, and elsewhere, and the big land grant universities such as Michigan and Wisconsin. But in the late twenties and thirties, professional geneticists distanced themselves from eugenics, while physicians began to embrace it.

Geneticists were beginning to despair of the excesses of eugenics as a movement. A group of ideologues had gone out of control, drunk on their own social and political success—and increasingly arrogant about flouting scientific standards. In 1924 the Johnson-Reed Immigration Act incorporated Harry Laughlin's eugenic advice and rolled back immigration quotas to 1890 levels in an effort to stem the tide of Asian and southern and eastern

European immigrants that the children of northern and western European immigrants found undesirable. A new wave of eugenic sterilization laws began to pass, many of them modeled on laws recently passed in Scandinavia. In 1927 the *Buck v. Bell* Supreme Court case upheld the constitutionality of Virginia's sterilization law and established the model for a statute that could not be overturned as unconstitutional. During the next fifteen years sterilization laws were implemented in some thirty-five states. Eugenics became accepted into popular culture. Fitter-families and better-babies contests at state fairs made the quest for perfection a friendly competition the entire family could enjoy. Laughlin grew reckless. His 1937 survey of the "human resources" of the state of Connecticut identified nearly twelve thousand people (roughly 1 percent of the population) as "definitely feeble-minded." He compiled a table correlating various crimes with intelligence quotient for inmates of the Connecticut State Prison for Men. He tabulated "crimes against chastity," "crimes against persons," "crimes against property," and "crimes against public policy," for persons with IQs from 0 to 125. But in his analysis, he lumped the upper middle of the distribution into a single group and failed to include the upper end in his statistics at all, thus flagrantly biasing his report toward the foregone conclusion. Laughlin was a crank. He caromed around the country like a 2 A.M. tippler, hurling insults and picking fights, giving free vent to blind prejudice, and dropping all but the pretense of scientific objectivity.[2]

If this was eugenics, serious geneticists wanted no more of it. Its utopian goals were farther from reach than they had initially thought. The ethical dilemmas were considerable—who decides what counts as "improvement"? And doing good, clean genetics on humans seemed almost impossible. Many genes affected a given trait; one gene could affect many traits; modifying genes distorted Mendelian ratios; spontaneous mutation and ordinary recessiveness meant that most traits never really disappeared. Humans were simply too difficult to work on, compared with crop plants or domesticated animals. The most rigorous, objective, and systematic record office in the world could not overcome the facts that humans are geographically promiscuous, that their chromosomes are tiny, numerous, and impervious to histological staining, and that ethical taboos prohibit controlled breeding experiments. As Lewellys Barker wrote in 1927, "Genetically speaking, humans are polyhybrid heterozygous bastards." Although most geneticists retained their wish to use genetic knowledge for human benefit, and a few attempted to reform eugenics to fit a more liberal, less coercive political agenda, most experimental geneticists at the time reluctantly tabled their vision of eugenic perfection until such a day when more reliable methods were available. Most geneticists who had dabbled in human genetics turned back to their more productive fruit flies, mice, and corn.[3]

A few true believers stuck to their principles. At the University of Chicago, H. H. Newman milked new insights from Galton's technique of studying twins to tease apart the effects of nature and nurture. Coming out of the psychological tradition of H. H. Goddard but lacking his or Laughlin's eugenic virulence, Newman's interest lay especially in understanding the genetic components of behavior and intelligence.[4] Many of those who sustained their eugenic vision were zoologists with an interest in heredity. At Wisconsin, Michael F. Guyer pursued a kind of human zoology, studying everything from human chromosomes to hereditary eye diseases to endocrinology, in addition to other mammals such as rabbits and rats.[5] Clarence Cook Little, a geneticist and the president of the University of Michigan, made the case for mammalian genetics to bridge the gap between the rigor and reliability of genetic knowledge in fruit flies and the profound social importance (and economic cost) of reducing the supply of mental and physiological defectives. Little's rhetoric occasionally blew a hose and overheated, but at bottom he understood that patience was the most important quality in a dedicated eugenicist:

> The rapid multiplication of mankind has forced the attention of man upon himself and his neighbors. His attention will remain there, his study be focussed there, his major problems will arise there, and there his chief success and failure will be experienced from the present time until a cataclysm reduces his numbers by hundreds of millions, or his control of his own physical and psychological progress, based on sound research, is determined and insured by inspired scientific leadership.[6]

Raymond Pearl took a different course. Rather than wistfully recognizing that eugenics was a lot harder than it looked, and rather than jettison his goal of doing productive human genetics, he clung to his hopes and abandoned the ship of agricultural science and orthodox eugenics. He sought instead a way of doing rigorous, valuable work on human heredity by exporting eugenics and biometry to public health and medicine. Pearl's work pioneered a shift in emphasis in eugenics toward population control. In his own mind, however, his main contributions were to a medical movement that, although minor, nevertheless enabled genetics to make a real incursion into the periphery of the clinical sphere.[7]

Pearl came out of the Galtonian tradition. Born in 1879 in Farmington, New Hampshire, he graduated from Dartmouth College in 1899 and in 1902 earned a Ph.D. in biology from the University of Michigan. He was a classmate of Herbert Spencer Jennings, another well-known geneticist with eugenic inclinations. In 1905 Pearl traveled to England to study statistics under Karl Pearson, Galton's protégé, a pioneer of biometry in his own right, and a passionate eugenicist. On his return, he took a position in agricultural breeding, studying poultry genetics at the Maine Agricultural Experiment Station. Pearl was big, both physically and intellectually; he was fearless, egotistical, witty, and an original thinker. All his life he relished stealing the spotlight and stirring controversy.[8]

A socially conscious, humanitarian man, he early on discovered eugenics as the science that claimed to apply genetic knowledge for direct human good. Like many conscientious geneticists of the day, he fantasized about human improvement. In a 1908 article titled "Breeding Better Men," he waxed fantastic about the potential for engineering life. "The time is rapidly approaching," he wrote, when it will be possible for someone to order up from a breeder a plant or animal with a specific set of desired characteristics, "just as though he were dealing with a manufacturing machinist." But more wonderful still will be the day when we can do the same with humans, for "from the human standpoint, whether individual or social, nothing compares in importance with the amelioration of man himself."[9]

In 1918 Pearl shifted from agriculture to medicine. William H. Welch hired him away from the Statistical Division of the U.S. Food Administration to join the faculty of his new School of Hygiene and Public Health at the Johns Hopkins School of Medicine in Baltimore. As a public health geneticist, Pearl had no patience for fussy studies of individual variation. He was antireductionist—he sought human improvement by the thousands, not one at a time. His interests were actuarial: birth, death, fertility, and

3.1 Raymond Pearl in the 1920s. Courtesy of Alan Mason Chesney Archives, Johns Hopkins University

longevity. But his interest spanned all levels, from social impact on down to physiology and genetics. He brought to bear on these problems any and all available tools, including statistics, genealogy, biometry, experimental breeding, racial pathology, and more. At Hopkins, he began to reject much of eugenic practice as dogmatic and unscientific, though he held fast to his vision of human improvement. Welch, a charter member of the American Breeders' Association and longtime supporter of various eugenic causes, must have been very pleased with the hire.[10]

At the Hopkins Hospital, Pearl founded what he called a Constitutional Clinic. He was, indeed, a major exponent of "constitutional medicine," a holistic approach to medicine that began about 1920, peaked in the thirties, and declined by 1950. Constitutionalism sought to revive and update the traditional medical concept of diathesis, most recently articulated by Archibald Garrod. The recent successes of bacteriology, they thought, had overshadowed constitution and predisposition. Why, the constitutionalists asked again, did one person get sick and the next not, when they were exposed to the same agent? They sought to better understand the soil on which the seeds of disease fell. There was an obvious connection between constitutionalism and heredity, and therefore between constitutionalism and eugenics. By the 1920s the constitutionalist movement was spearheaded by such scientifically oriented physicians as Baltimore's Lewellys Barker, Walter Alvarez from Berkeley and then Minnesota, and the Rockefeller's George Draper, as well as medically oriented biologists such as Pearl and Charles Stockard, of Cornell University Medical College. Nearly all of them supported eugenic principles of human biological control and improvement.[11]

During constitutionalism's heyday in the twenties and thirties, its authors emphasized the uniqueness of the individual as a constellation of heredity, experience, and activity. Draper, the figurehead of the movement with his Constitution Clinic in New York, developed a system of "panels" or domains of health and personality—morphological, physiological, psychological, and (farsightedly) immunological—which he envisioned every person to uniquely manifest.[12] Constitutionalists tended to be more interested in classifying disease than in classifying people. This, perhaps, is why the Scandinavian physician Knud Faber concluded his 1923 book *Nosography* (an ancient term referring to the classification of disease) with a chapter on constitutional pathology. Although doctors had long recognized heredity as playing a role in disease, he wrote, it had been insignificant within the scope of disease as a whole. The greatest killers, after all, had been infectious diseases. However, "an enormous change" had recently occurred, as a result of recent advances both in bacterial pathology and in genetics. Heredity had lately been recognized as playing a profound role in disease, Faber wrote. He surveyed knowledgeably the numerous human diseases then known to follow Mendelian patterns. But he criticized the parsing of disease into endogenous and exogenous—loosely, "internal" diseases, largely hereditary, and "external" or infectious, toxic, and so forth. He saw constitutionalism as dissolving that artificial barrier. "What is a cause?" Faber wrote.

> Is it justifiable to say that a disease has one cause? Has not disease, like all other phenomena of this world, not only one, but many causes, and is it not more correct to say that the morbid phenomena in each separate case appear as a result of a series of conditions? Indeed, ought we not give up speaking about "causes" in pathology, as we have done in the other physical sciences and be content with speaking of conditions?[13]

Constitutionalists such as Draper sought to understand the multiple causes in disease and in particular to grasp the role of idiosyncrasy.[14] Why did this person have this disease at this time? In Pearl's journal in 1929, Draper wrote of disease as not an entity but a maladjustment, "a subtly moving, changing set of reactions between man and his environment." Garrod himself acknowledged Draper's Garrodian lineage in his 1931 book *Inborn Factors in Disease.* However, Draper was also cozy with the Galtonian camp. He was elected to the Galton Society in 1923, and was close with Davenport and Laughlin and later joined the Eugenics Research Association of Cold Spring Harbor. They apparently found it possible to overlook Draper's relative lack of genetic determinism; more important, it seems, was the common ground between them on such topics as racial classification. When Draper retired, he was replaced by the far more dogmatic William Sheldon. His version of constitutional medicine incorporated his "somatotype" theory, an anthropometric method of classifying humans by physical characteristics and inferring from these measures nearly everything about one's health and personality. Soon Sheldon had steered constitutional medicine toward a typological, anthropological approach that had more in common with Henry Goddard than with Archibald Garrod, which doubtless hastened its decline.[15]

Pearl developed his own idiosyncratic style of constitutionalism. In 1925 he became the director of a new Institute of Biological Research, a multidisciplinary research center administered by the university, designed to probe the constitutional (largely genetic) and environmental factors in disease. Within the Institute, Pearl designated a Division of Medical Genetics, the first instance of that name in an American medical school. The two bodies were distinct only administratively—they occupied the same space and had the same staff. Pearl's group studied the hereditary basis of complex traits such as tuberculosis and hypertension, which had long been recognized to have a hereditary component in addition to the obvious environmental aspects. They collected data and performed analysis on many other diseases as well. Health-minded philanthropies funded the work: the National Tuberculosis Association, the Russell Sage Foundation, the Commonwealth Fund; after a year, Pearl presented a proposal to the Rockefeller Foundation, which provided a major multiyear grant. As genetics moved into medicine, the links to Progressive-era health reform held fast.[16]

The work was a novel hybrid of biometry, genetics, and medicine, focused on what are still considered some of the thorniest problems in medical genetics. Two primary concentrations were hereditary predisposition to tuberculosis and genetic factors in hypertension. Pearl the statistician was relatively uninterested in Mendelian genetic diseases, which are relatively rare. He wanted to address medical problems of major import. One of his biggest projects was a multiyear study of the constitutional factors contributing to length of life. Decades later, the Johns Hopkins cardiologist-turned-geneticist Victor McKusick would hire back Pearl's assistant Blanche Pooler to help him analyze and extend Pearl's findings. The genetic basis of longevity is a vigorous, well-funded area of research today, and is even integral to such biomedical fringe movements as transhumanism. Like intelligence, antisocial behavior, and predisposition to disease, longevity—or life extension, as Irving Fisher had it—is one of the fundamental themes of medical genetics, from the eugenics era to the genome age.[17]

Pearl ran his operation in a manner consistent with its hybrid structure, as a combination of Eugenics Record Office and medical clinic. He hired fieldworkers and assembled pedigrees and Galtonian records, just as Davenport did at Cold Spring Harbor. Pearl's term for these was Family History Records. Pearl shelved many of Davenport's categories—the occupations and many of the dimensions of physiognomy—and added a number of more clinical measurements and descriptions to the examinations. Still, the family resemblance between Pearl's records and Davenport's is strong. Pearl's "individual records" constituted a multipage Galtonian questionnaire, with name, address, sex, marital status, race, occupational history, clinical history, medical features, pregnancies, skin conditions, and so forth.

There was a page for laboratory studies, such as kidney function, urinalysis, and the Wasserman test for syphilis. A good disciple of Pearson, Pearl conducted a lengthy biometric examination, including height, weight, handedness, various measures of the head and other body parts, and physical description of everything from hair to nails. Another section of the record form covered "home environment," including size and description of the home, how rooms were used, how clean the home was, family income, existence of cousin marriages, and characteristic diseases of the family. Like most Progressive-era eugenicists, Pearl was deeply interested in race. He employed a lengthy and telling racial taxonomy that included five distinct categories for Scandinavians and four for Britons, but four also for all of Latin America, three for all of Asia, and one for "Negro." Pearl's science was quantitative, systematic, as objective as he could make it—and yet it was profoundly colored by contemporary social norms and prejudices.[18]

In 1924 Pearl collected a number of his essays into a volume under the title *Studies in Human Biology.* He intended the term to express his synthetic yet deterministic philosophy that all studies of humans, from economics and politics on down, were ultimately grounded in biology. In 1929 he founded a journal under the same name, which he intended to serve as a forum for any and all studies of the human animal. He assembled an appropriately wide-ranging editorial board that spanned the globe and the range of knowledge. Among them were Charles Davenport from Cold Spring Harbor, Aleš Hrdlička, a eugenically minded anthropologist from the Smithsonian Institution, Clark Wissler, a fellow member of the Galton Society, the German eugenicists Eugen Fischer and Ernst Kretschmer, H. Lundborg from Uppsala's State Institute for Race Biology, the behaviorist John B. Watson, and the economist Bronislaw Malinowski. The inaugural issue looked a lot like a eugenics journal. It featured articles by Davenport on racial differences in mental capacity, by Henry Fairfield Osborn on anthropology, and by C. Todd Wingate on "entrenched Negro features." It led off with a long article by Pearl himself outlining the methods of the Constitutional Clinic and Division of Medical Genetics.

Pearl thus remained a strong, even passionate supporter of the Galtonian eugenic perspective. He never doubted that race was deeply significant as a marker of fundamental human characteristics, and he sought to use biological, especially genetic, knowledge to improve health, intelligence, and well-being in the long term. Yet by the late 1920s, the outspoken Pearl became a scathing critic of the Eugenics movement, which he capitalized to distinguish it from the general principles of human genetic improvement. He showed a bit of the flair of his friend H. L. Mencken in some of his searing critiques, such as a 1927 article in Mencken's *American Mercury*, in which he called Eugenics "a mingled mess of ill-grounded and uncritical sociology, economics, anthropology, and politics, full of emotional appeals to class and race prejudices, solemnly put forth as science, and unfortunately accepted as such by the general public." Since he himself was mingling the human sciences, probing racial differences, and publishing it all under the heading of human biology, Pearl's critique here is the mess, the prejudice, and the sanctimony of eugenic ideologues, not the principles of eugenics itself. Eugenics was not doing the wrong thing—it was doing the right thing badly. By importing eugenic questions, methods, and even personnel into the medical sphere, Pearl sought to rationalize, systematize, and legitimize the grandly synthetic, humanitarian ideology that to him eugenics represented. He intended human biology as eugenics done right, with good statistics, rigorous standards of objectivity, a broad-minded liberal social agenda, and medical affiliations.[19]

Pearl talked a better game than he played. After five years, his funding dried up and the Institute closed. Still, he left his imprint on biomedicine. Articles on "human biology" crop up regularly in the medical literature after 1930, and the term still enjoys some use today. Further, both his methods and his data return in the subsequent development of American medical genetics. Pearl's Institute should be considered a false start among genetic approaches to medicine, but one with important foreshadowings of future trends.

Meanwhile, however, Pearl and his Hopkins colleague Adolf Meyer helped carry out an experiment in practical eugenic medicine. From 1927 to 1932 they were central within the organization of Baltimore's Bureau for Contraceptive Advice, one of the first open birth control clinics in the country. Birth control laws were liberalizing rapidly, much to the delight of the stalwart activist Margaret Sanger, who was still active in the American Birth Control League. True to the biomedical principles of its founders, the

Bureau for Contraceptive Advice was both a research facility studying the social impact and demography of the dissemination of birth control information and a clinical facility actually offering that information. Both Pearl and William H. Howell, another director and the dean of Hopkins's School of Public Health and Hygiene, were avid birth control advocates, but wanted to avoid Margaret Sanger's inflammatory politics and rhetoric and to offer instead a more conservative and scientific approach to birth control. Meyer, who had been involved in the birth control movement for years and served on the birth control committee of the American Eugenics Society, was a natural and enthusiastic adviser to the project. Although the board also included the obstetrician J. Whittredge Williams, the Bureau's second annual report stipulates explicitly that no advice was given to women already pregnant—this was not, he wished to assure his supporters, an abortion clinic.

In 1932, after the fourth year, Pearl, the Bureau's statistician, delivered a statistical report on the results of the experiment in social medicine. He described the collection of biometric data, which was a version of his own work with the Constitutional Clinic and Division of Medical Genetics. Clinical histories were "recorded with extreme care," he noted—in contrast to his opinion of much eugenic data. Such histories, he noted, "contain a wealth of material for the study of various medical and biological problems." The Bureau reflected the class and race politics of sex in the 1930s. "Most of our patients are dispensary type," that is, poor and unable to pay for medical care. "About one fifth have been colored." White patients were seen on Wednesday afternoons, "colored" patients on Thursday afternoons. After a thousand cases, the research project was shut down and the Bureau was converted to a purely clinical, community-oriented organization, the Baltimore Birth Control Clinic, "for the benefit of the women of Baltimore." In 1938 it joined Sanger's Birth Control Federation, which in 1942 merged into Planned Parenthood, the principal American birth control advocacy group today.[20]

Constitutional medicine and human biology illustrate the complex interplay between what I have called the Garrodian and Galtonian strains in medical heredity. Pearl was an early eugenics enthusiast who eventually rejected the movement in favor of doing eugenics properly and in a medical public health context. Both he and Draper maintained professional relationships with eugenic leaders well into the 1930s. Both employed classical Galtonian anthropometric methods. Both imported those methods and ideas into the medical clinic. It was a time of intense race consciousness, and neither Draper nor Pearl seems to have questioned conventional racial stereotypes any harder than those eugenicists they criticized. Yet philosophically they were constitutionalists, Garrodian at the core. Their interest in heredity was in how it shaped our responses to the environment. They were holists and synthesizers, interested in unifying heredity with other branches of knowledge. In order to make sense of the baffling complexity of the total human being, they resorted to typologies—and often standard ones, unreflectively drawn from the surrounding culture. This brought the constitutionalists at times uncomfortably close to those eugenicists they railed against. It was, in the end, not very productive. Given the tools of the day, this purely populational form of constitutionalism lacked explanatory power. For all their power for generalization, statistics were insufficient to tease out the filaments of human constitution; they omit mechanism. Draper and Pearl were deterministic and antireductionist at a time when medicine was increasingly reductionist and antideterministic. The "soil" remained refractory.[21]

* * *

A contrasting approach was to treat genes as "seed"—as agents of disease. William Allan, the southern country doctor, was trained in germ-theory medicine. Born in 1881 in Catonsville, Maryland, near Baltimore, Allan took an M.D. in 1906 from Baltimore's College of Physicians and Surgeons. His first appointment was in Charlotte, North Carolina, where he was associate professor of bacteriology and director of the clinical bacteriology laboratory at the North Carolina Medical College, chartered in 1893 by Davidson College.[22] His early research interest was amoebic dysentery.[23]

After the war, his interests began to shift. His patients led him to genetics—and to the application of another principal technique of human genetics: the isolated population. The taboo on human breeding experiments forces the human-geneticist to seek "natural experiments," where people voluntarily mate in interesting ways. Charlotte, located in the

3.2 William Allan in 1941, at the time of the founding of the first Department of Medical Genetics in the country. Courtesy of Wake Forest School of Medicine

Appalachian Piedmont, was a natural laboratory for human disease genetics. A close-knit family structure, very low rates of mobility, and relatively high rates of inbreeding uncovered much latent genetic disease. Family pedigrees could be assembled for particular traits, enabling the geneticist to determine the manner of inheritance (dominant, recessive, sex-linked, polygenic, and so forth) and perhaps to correlate the trait with other characters. Although Allan had little or no formal training in genetics, he grasped intuitively that such an isolated population was a valuable natural experiment for the study of pathological heredity. He could see patterns, but he lacked the statistical training to analyze them.

Perhaps through word of mouth, Allan learned of the young Harvard geneticist on the faculty in Raleigh. Laurence Snyder was a man of intelligence and immense confidence, with a flair for writing, teaching, and administration. Later photos show a bald, stern man; biographies record his fondness for cigars and his talent for boogie-woogie piano.[24]

Snyder's interest was blood groups—the only productive model system in human genetics at the time. "Blood," of course, has deep folk roots as a signifier of heredity—especially race—despite studies dating back to Galton that showed that the hereditary elements did not circulate freely. Blood groups are just as old as Mendelian genetics, and the two histories are deeply intertwined. The same year that de Vries and Correns rediscovered

3.3 Laurence H. Snyder in 1935, when he was teaching the first genetics course in an American medical school, at Ohio State University. Courtesy of Ohio State University Archives

Mendel's work, Karl Landsteiner found that humans differed in the pattern of agglutination of their red blood cells. Serologists could distinguish one blood type from another and recognized immediately the potential benefits for legal cases, clinical medicine, and, in the First World War, military medicine. In 1922 R. Ottenberg advocated the use of blood group data in cases of disputed paternity. J. A. Buchanan and Frederick Reichert criticized him, however, arguing that the genetics was too primitive for the data to be reliable. "It appears," they wrote, that "conclusions have been arrived at, and applications made, without sufficient investigation." They continued, "There still remains a need for a thorough and extensive field study, carried beyond two generations. Then, with the data findings, a more authoritative and convincing hypothesis as to the mechanism of the inheritance of blood grouping could be advanced."[25]

The test for blood groups was so easy and had such immediate medical relevance that hospitals soon began performing it routinely. This led to huge amounts of clean, reliable data, on a scale far beyond anything collected by the Eugenics Record Office or the Galton Laboratory. However, no simple Mendelian explanation fit the pedigrees. In 1919 the Polish bacteriologist Ludwik Hirszfeld and his wife, the pediatrician Hannah Hirszfeld, explained the blood-group trait as the result of two Mendelian genes, A and B. In 1924, with the benefit of much more data that had been collected in the interval, Felix Bernstein argued that the blood group determiners resided at a single locus, with three alternative alleles, A, B, and O. The debate raged over the rest of the decade.[26]

When Bernstein's paper appeared, Snyder was a graduate student at Harvard's Bussey Institution under William E. Castle. Castle, recall, was an agriculturally oriented geneticist with a deep and long-standing interest in eugenics. Probably through Castle's mediation, Snyder had spent time under Davenport at Cold Spring Harbor and had developed a strong interest in both mathematical human genetics and eugenics. The blood group problem appealed. He took up the torch for Bernstein and became, in Pauline Mazumdar's phrase, "Bernstein's bulldog." His 1925 thesis, published in the *American Journal of Physical Anthropology,* applied the triple allele hypothesis to the characterization of racial groups. Human blood groups, he wrote, "occur as fixed bio-chemical conditions, subject to the laws of heredity. As such they provide a method of studying racial origins and relationships." He divided humans into six racial groups—European, Hunan, Indomanchurian, Africo-Malaysian, Pacific-American, Australian—according to the proportions of the four blood groups they contained. Snyder believed resolutely in a single origin for the human species, with subsequent radiation into the different groups, a "monogenist" view, in contrast to the Hirszfelds' polygenist interpretation. The seemingly dry debate over monogeny vs. polygeny had ideological overtones: under a polygenist interpretation, it was possible to argue that, for example, blacks and whites were different species and therefore that miscegenation was bestiality. Snyder's work was heavily racialized, and he sometimes advocated quite harsh eugenic measures, but he never saw the human race as composed of multiple species.[27]

Blood group genetics developed rapidly in the 1920s—faster in Europe than in the United States—and it was related at least as closely to anthropology as to medicine. In 1927 Otto Reche, a professor of anthropology and ethnography at the University of Vienna, issued a call for a new society for the study of blood groups. Also that year, blood group researchers gathered at the annual meeting of the Institut International d'Anthropologie, organized by the Ecole d'Anthropologie de Paris. Ludwik Hirszfeld, Snyder, and other leading blood group researchers attended. In 1928 a new institute of blood grouping was formed in Charkow, Ukraine, which boasted the membership of the global leaders of blood group research, including Landsteiner, Ludwik Hirszfeld, Snyder, Reche, and others, and which published a new journal of blood grouping in German. By the middle of 1928 a *Eugenical News* editor could write, "No other branch of human heredity is so well organized as the study of the blood groups."[28]

In the United States, Snyder led the way in importing blood group genetics from anthropology into medicine. In a long review in the *Journal of Laboratory and Clinical Medicine* in 1927, he discussed numerous applications of blood group genetics, including race classification, blood transfusion, and disputed paternity. He expanded these ideas into a book, *Blood Grouping in Relation to Clinical and Legal Medicine,* in 1929. He found himself drawn increasingly to the medical applications of blood grouping and soon of human genetics in general.[29]

* * *

Through the North Carolina summer of 1928, Snyder tagged along with Allan to Appalachian church socials and family reunions, collecting pedigrees and blood samples. Their collaboration fused mathematical population genetics, ERO-style fieldwork, the genetics of isolated populations, and clinical medicine. This approach provided a rare foothold on the problems of how to do legitimate human genetics and accomplish social good with heredity. Snyder and Allan remained close colleagues and friends until Allan's death in 1943.

Their styles were complementary, and their collaboration stimulated each man to evolve both methodologically and philosophically. While Allan's influence pushed Snyder from anthropology into medicine, Snyder helped Allan sharpen his nebulous and largely unproductive conception of heredity into a more rigorous Mendelian framework. Allan's earliest genetic studies explored complex and often subjective diseases such as headache, hypertension, and diabetes. By the mid-1930s, he began to examine diseases with less clinical importance but clearer heredity. In a paper on inherited night-blindness (retinitis pigmentosa), he found that a dominant form of the disease tends to be less severe and occurs later, a recessive form sets on early and leads to total blindness, and a sex-linked form is intermediate between the two. It is an early example of the dissection of one disease into several based on correlations between hereditary pattern and etiology—an important technique of medical genetics in the genome age. Allan also examined xeroderma pigmentosum, peroneal atrophy (degeneration of muscles of the lower leg, a condition now recognized as part of the Charcot-Marie-Tooth syndrome), and muscular dystrophy—relatively rare conditions with clear Mendelian patterns, and all still of interest to genetic medicine today.[30]

Not long after his collaboration with Allan, Snyder developed the first genetic test. Knowledge of human genes remained scant. Colorblindness and hemophilia had been mapped to the X chromosome by pedigree analysis. Somewhere on the nonsex chromosomes were two blood groups—the ABO blood group and the MN blood type, discovered by Landsteiner and Philip Levene in 1927. Other Mendelian traits included eye color; the direction of the whorl of hair at the back of the head; the presence of hair on the second joint of the fingers and toes; and William Allan's dominant form of migraine. Six in total. It was therefore a real event when, in 1931, Snyder found a seventh—and suggested a simple, cheap, and portable way to test for it.[31]

Phenylthiocarbamide, PTC for short, is a sulfur-based compound, used in the rubber industry. Related compounds give the bitterness to broccoli and Brussels sprouts. Snyder read a one-paragraph note in *Science* by Arthur L. Fox, an industrial scientist at Du Pont in Wilmington, Delaware, who described "tasteblindness" to PTC: to most people, it tastes extremely bitter, but a significant fraction cannot taste it at all. Snyder read the note and immediately wrote to Fox for samples. At Cold Spring Harbor, Albert F. Blakeslee also became interested in the PTC story. He and Snyder worked independently to confirm whether the ability to taste PTC is indeed a Mendelian trait, if so, how it segregates, and its biochemical and genetic nature. It behaved like a classic dominant, with both alleles (taster and nontaster) being common.[32]

By soaking a slip of paper in a solution of PTC and letting it dry, a researcher had a quick test for the presence of the PTC taster allele, a test equally as reliable in the field as in the lab. Within a year, Snyder and others were testing patients with other heritable conditions for PTC sensitivity in the hope that the two traits would be inherited together. It was an utter stab in the dark, but it was the only way to develop so-called linkage groups. In maize by this time, there were ten big linkage groups, and the Cornell cytogeneticist Barbara McClintock had correlated each one with one of the ten chromosomes. In *Drosophila*, Theophilus Painter had recently discovered the giant "lampbrush" chromosomes of the salivary glands; Calvin Bridges was feverishly mapping the hundreds of known fly genes to precisely defined locations on each of the chromosomes. In contrast, human geneticists were groping for any linkage at all. Snyder's discovery was a tiny step, but at last human genetics seemed to be getting a bit of much-needed traction.[33]

Snyder's other main contribution was "Snyder ratios," a technique for estimating the frequency of heterozygotes. This was one of the basic problems of human genetics. Think of a basic Mendelian trait, such as brown (B) vs. blue (b) eyes. Blue-eyed people must have two copies of the recessive allele: bb. But brown-eyed people can be either BB or Bb. How do you tell the heterozygotes (Bb) from the dominant homozygotes (BB)? The question is fundamental to genetic analysis and has great clinical impact. Often, disease alleles are recessive, meaning that heterozygotes have no symptoms but carry the disease—heterozygotes were analogous to "carriers" of infectious disease. Plant and animal geneticists uncovered carriers by mating individuals back to a known recessive—one of the parents, in a classic Mendelian experiment. A principal workaround for this problem was to estimate the number of heterozygotes in a population. Snyder found that the number of heterozygotes expected from matings in which one or both parents was dominant was a function of recessive gene frequency.

However, Snyder's biggest contributions to the field were not intellectual. In the thirties and forties he emerged as one of the leading forces in professionalizing human and medical genetics. He designed and taught the first genetics course to medical students in the country. He established four medical genetics programs on his own and was directly involved in founding at least two others. He was a founder of the field's first professional society in 1948. He gave lectures and wrote articles promoting the field as a vital dimension of medicine. Like Victor McKusick a generation later (see chapter 6), Snyder is a founding father of American medical genetics mainly because of his professional and disciplinary contributions.

* * *

The founding mother is Madge Macklin, an American physician who worked in southern Canada. In the 1930s Macklin was the nation's most outspoken advocate for medical genetics. She articulated an impassioned, closely reasoned argument for the inclusion of genetics in both medical education and medical practice. Aided by Snyder and Allan, she shaped what became the profession of medical genetics, which in turn was the incubator for modern genetic medicine. Macklin's vision of medical genetics, then, is woven into the fabric of American biomedicine.

She was born Madge DeGrofft Thurlow in 1893, in Philadelphia.[34] Her family soon moved to Baltimore, where she grew up and attended the all-girls Western High School, not ten miles from William Allan's home. In 1914 she earned her bachelor's degree from nearby Goucher College, at that time a women's school. In 1915 she won a fellowship to Johns Hopkins Medical School. In 1918 she married Charles C. Macklin, an associate professor of anatomy at Hopkins. Madge Thurlow Macklin took her M.D. the following year. In 1921 the Macklins moved to the University of Western Ontario, in London. Charles took a tenured position in the Department of Histology and Embryology, while Madge was appointed part-time instructor. She never rose beyond assistant professor; reportedly, she clashed with the university administration. In 1930 she helped found the Canadian Eugenics Society, serving on its executive committee from 1932 to 1935 and in 1935 as

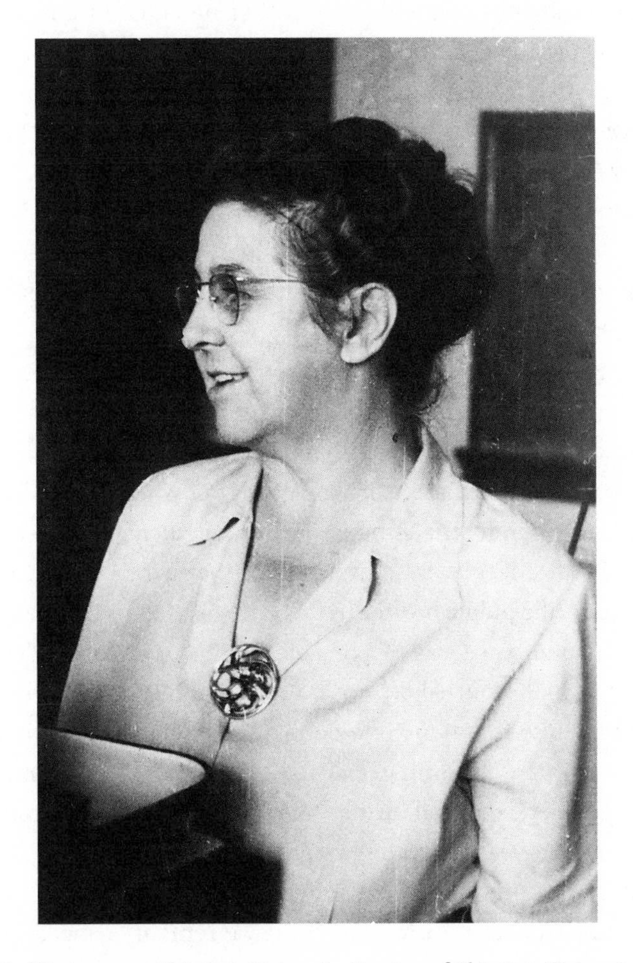

3.4 Madge Macklin in 1955, at Ohio State University. Courtesy of Ohio State University Archives

its director. She was one of the few medical doctors involved in the American Society of Human Genetics in its early years, and she served as its president in 1958. She was small—five-foot-two—energetic, and, by middle age, obese. She had a passionate concern for social justice and yet, says the historian of Canadian eugenics Angus McLaren, she lacked sympathy for the erring and could be combative and intolerant. She became a fierce partisan for using the physician's authority for social amelioration of hereditary ills.[35]

Macklin, Allan, and Snyder mounted an argument for medical genetics drawn from population studies. Their strategy was to analogize "horizontal" (infectious) transmission with "vertical" or hereditary transmission. For Macklin, germs and genes were parallel disease agents. "It makes no difference whether the disease with which they are afflicted came to them by way of germs or germ plasm," Macklin wrote; "the disease is transmissible, and so comes under the category of public health." Allan agreed. "The public health movement in this country has always travelled along a single track," he wrote, "leading to the control of infectious disease. It seems to me high time that we opened another door in the public health field leading to the control of hereditary disease."[36]

This small campaign to bring genetics into medicine coincided with a general increase in interest in chronic disease. By the 1920s cancer and heart disease had replaced tuberculosis, cholera, and diphtheria as the nation's top killers. Historians have argued over whether the change came willy-nilly out of industrializing society or whether public health reform played a significant role. But physicians of the 1930s believed it to be the result of scientific public health. In their minds, not just sanitation but the germ theory of disease was responsible for this dramatic epidemiological shift. Writing in the journal *Science* in 1934, John Murlin, a professor of vital economics, wrote that if anyone doubted the advances of science in the twentieth century, "let him recall the typhoid fever epidemics of only 30 years ago." Let him recall further "the enormous mortality from tuberculosis only twenty years ago; the ravages of rickets in children of the tenements only ten years ago—all these and many other diseases either wholly brought under control, to remain so (if we remain civilized), or rapidly yielding to the science of prevention and treatment."[37]

Not dying of tuberculosis at age twenty-five meant surviving to contract cancer at fifty; the falling tide of acute disease was revealing tidepools of chronic disease (fig. 3.5). Genetics, Macklin wrote in 1932, was "becoming increasingly important as the death rate from constitutional disorders mounts, due to the decrease in death rate from infectious diseases." Allan, too, wrote of the "remarkable shift during the past thirty years in the chief causes of sickness and death in the general population, from infectious diseases to chronic constitutional or hereditary diseases." And in another article: "Our most serious problems in medicine and public health have to do with those so-called constitutional diseases that are now the leading causes of death."[38]

The main medical benefit of genetics lay in prevention. In most cases, no therapeutics were available, but steps could be taken to reduce incidence or even eliminate the disease altogether. One of Macklin's first papers dealing with heredity, a 1926 article cowritten with her Western Ontario colleague J. Thornley Bowman, showed that peroneal atrophy was passed down as a single-gene Mendelian dominant. In a final section of the paper, "Eugenic

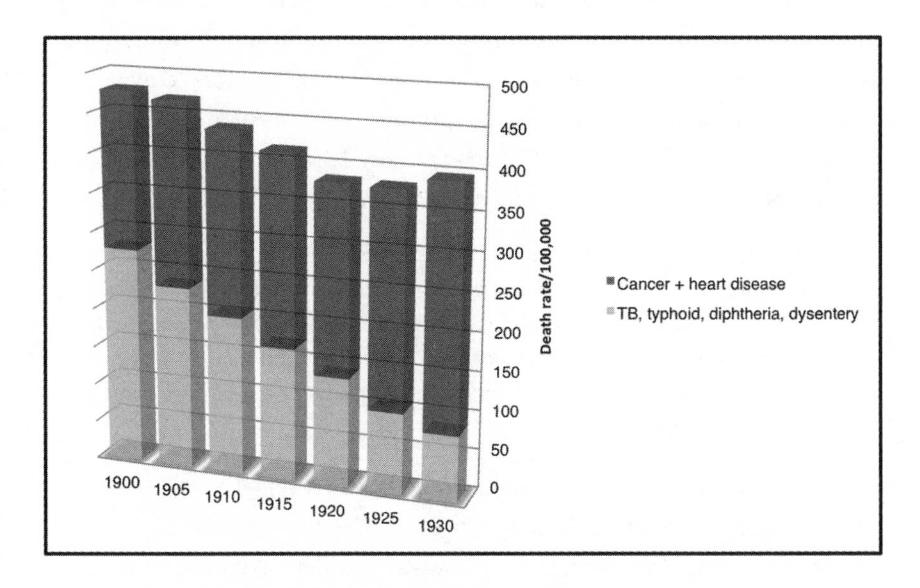

3.5 Declining rates of infectious disease and climbing rates of chronic and hereditary disease. Compiled from Forrest E. Linder and Robert D. Grove, *Vital Statistics Rates in the United States, 1900–1940* (Washington, DC: Government Printing Office, 1947), table 20, pp. 330–32.

considerations," the authors wrote, "It is obvious that a disease such as peroneal atrophy should be eliminated from society, if possible. Ways of accomplishing this at once suggest themselves. First of all, we would place the value of thoroughly educating the affected as to the certainty of their potency to transmit the disease. Continence, contraception and even sterilization might be enjoined." In a paper on the genetics of eye diseases the following year, Macklin made a similar claim: "It is at once obvious that such a disabling condition as glaucoma should not be increased in the community." A unit character should be treated as a disease agent, like a germ, and efforts made to eradicate it.[39]

Macklin, Allan, and Snyder were not coy about what the elimination of genetic disease implied. "This particular field of public health is really eugenics," said Macklin.[40] By this she meant negative eugenics, the effort to reduce the numbers of those deemed unfit. In an article titled, "Medicine's Need of Eugenics," Allan brought in classical imagery and what seemed to him common sense. "In the days when Sparta flourished, defective children were exposed on the mountainside to die," he wrote. "In this Christian era defectives are carefully nurtured, and if their defects are hereditary are allowed to pass them down to future generations. Would it not be more sensible in the case of hereditary defects that are predictable to avoid the production of such children?" Like Galton before him, Allan saw eugenics as an alternative to heartless Spencerian social Darwinism. In the 1940 edition of his textbook, Snyder let his rhetoric soar:

> It is the hope of the student of human genetics that such a cooperative line of research may eventually give rise to a social edifice, the foundation of which is made up of substantiated facts about the development, both from a genetic and an environmental standpoint, of human characteristics, and the superstructure of which is a tower of eugenic strength which can be defended against any attack.[41]

Snyder, Macklin, and Allan articulated a vision of Mendelian public health in which genes were like germs, agents of disease that needed to be sequestered and prevented from spreading. It was self-evident to them that parents with known hereditary diseases should not have children; they employed the physician's reasoning of the individual as living within a community and therefore bearing a part of the responsibility of the community's overall health.

Under a germ theory of genes, the equivalent of quarantine is sterilization. In 1932 Macklin rationalized the most invasive procedure of eugenics on medical grounds, by evoking terrifying hordes of genetic Mary Mallons that only the genetically knowledgeable doctor could stop:

> If we make a typhoid carrier give up her job as cook, isolate her, capture her when she escapes, and keep her practically a prisoner until she is rendered incapable of transmitting her disease, are we not justified in isolating and rendering incapable of transmission a person who infects not a few with whom he comes in contact but the whole future race descended from him?

The argument plays more on fear than on math. The notion of a genetic carrier implies a recessive allele that would be deleterious only if one of those future descendants married another carrier, and even then there would only be a one in four chance of the disease reappearing. Even if she could be identified with certainty, a genetic Typhoid Mary seems a much smaller threat than the original. The alarmist notion of a genetic epidemic has echoes of Progressive fear-mongering over "degeneration." The germ theory of genes could be misleading.[42]

The way to spread the gospel most effectively was in the classrooms of the nation's medical schools. Macklin was particularly effective in pushing for the inclusion of genetics in the medical curriculum. In several articles published in 1932 and 1933, Macklin named the new medical specialty. She admitted, though, that genetics was a young science, and that "genetics as applied to medicine is a mere infant. But it is a very lusty one and will keep on crying until it is heard." A residue of the old spirit of progress kept her from faltering because of ignorance. The fact that little was known about the genetics of disease ought not to stop physicians from applying what they do know. "Although we cannot speak with that precision upon many of the topics related to it that we would like," she wrote in 1932, "we will make progress only as we acquaint our medical students with what we know." Snyder took the point. "Progress is being made rapidly at present and will be made even more rapidly in the future," he wrote. "It is therefore of advantage to the medical student to become familiar with the present knowledge of human heredity, incomplete though it may be, in order to be able to evaluate the statements and articles of the future as they may appear."[43]

Medical genetics benefited medicine on several levels, Macklin wrote. In 1912 Davenport had argued that physicians needed to know about genetics for reasons ranging from the individual to society. Macklin now did the same for the teaching of genetics to medical students. First, she said, it would improve diagnosis. A course in medical genetics would acquaint the student with rare diseases he would not otherwise see. Some of these are easily confused with nonhereditary diseases but require different treatment. She cited an example of an apparent case of hypothyroidism. Thyroid medication, the usual therapy, made this patient feel worse. A consulting physician recognized his condition as ectodermal dysplasia, a hereditary disease. He discontinued the medication and the patient improved. The pedigree is simply another diagnostic tool—the hereditary basis of the disease made no difference to his prescription. Therapy has never been the strong suit of medical genetics, but Macklin argued that knowledge of heredity made for more rational therapy. She argued that understanding the hereditary principles underlying disease would lead to better, more accurate, and more appropriate therapeutics—a familiar promise to anyone who reads the biomedical news today.[44]

Macklin understood that pedagogy was the best vehicle of recruitment. "Only as we teach the students a course in pathological heredity"—a synonym for medical genetics—"will we have teachers trained to bring [genetic] illustrations into their teaching." The course, she continued, would be a "great asset in preventive medicine and public health." She published a draft syllabus for a course in medical genetics, which she hoped would be adopted in medical schools across the United States and Canada. She envisioned a thirty-six-lecture-hour course, to be taught in the fourth year of medical school. It would devote two hours to the basic principles of Mendelian genetics, followed by a series of lectures on genetics as applied to various categories of disease: endocrine, blood, digestive, metabolic, eye, skeletal, and so forth. Five hours would be devoted to diseases of the nervous and neuromuscular systems, and four more to mental diseases. The last hour was to be devoted to "preventive and public health aspects of the problem," or eugenics.[45]

The next year, Snyder offered a course developed "in full sympathy with the views of Dr. Macklin"—the nation's first course in genetics offered as part of the regular medical curriculum. The course initially took as its text *Human Heredity*, by Erwin Baur, Eugen Fischer, and Fritz Lenz. This was the third edition of a now notorious textbook that first appeared in 1921, under the title *Outline of Human Heredity and Racial Hygiene*. Highly praised by Hitler, Baur-Fischer-Lenz served as the handbook to the Nazi eugenics and euthanasia program as well as becoming the standard text on human genetics in the 1930s in Germany, America, and elsewhere. Snyder preferred Baur, Fischer, and Lenz over the somewhat older text, *Genetics and Eugenics*, written by his mentor, William E. Castle.[46]

In 1935 Snyder published his own textbook, *Principles of Heredity*, which went through five editions and was widely adopted in genetics courses. He covered, of course, the rudiments of Mendelism and basic population genetics. Though it was a general genetics text, Snyder used human examples wherever possible. The last four chapters cover humans—inheritance of physical and physiological traits, inheritance of mental traits, eugenics, and the analysis of human family histories. Snyder dedicated the book to Charles Davenport.[47]

The early chapters of *Principles of Heredity* indicate that Snyder was moving toward an increasingly subtle understanding of gene action and the relation of genes to environment. In 1933 he had written that "heredity and environment are cooperative in the production of any finished character," and his textbook elaborated that view. He dealt with the mechanics of inheritance, including Mendelian and non-Mendelian patterns, cytology, and population genetics. He devoted a chapter to multigene traits, which complicate the genetic picture immensely. A chapter titled "How Genes Act" included a proposed genetic explanation of embryological development and biochemical genetics. Much of the book demonstrates that Snyder was indeed well up on the current genetic literature and debates, particularly the emerging understanding of genetics as a highly complex process, contingent on the interactions of many genes with one another and with the environment. When dealing with genetics, Snyder could be a subtle thinker, neither dogmatic nor rigidly hereditarian.[48]

But when he wrote about eugenics, nuance abandoned him. In the penultimate chapter, "Eugenics," retained in the 1940 edition, he wrote that although we may never agree on an ideal human genetic constitution, probably everyone could agree that some traits "are so undesirable that the race could well do without them." Much of what follows reads like standard eugenics, mixed with the occasional reference to recent biochemical work, such as work by Asbjørn Følling and others on phenylketonuria. First among the traits to eliminate is "feeble-mindedness," that shibboleth of classical eugenics; Snyder calls it "probably the outstanding problem in eugenics." He cites estimates from Paul Popenoe's Human Betterment Foundation that six million Americans—one in twenty—were feeble-minded. The page opposite carries an illustration, credited to the *Eugenical News,* of an entire family with achiropodia, a grotesque condition which leaves its bearer with truncated, skin-covered sticks instead of forearms and lower legs. A horrific collection of microcephalics, cretins, and Mongolian idiots populates the pages, much as clinical illustrations of birth defects fill medical texts. They leave no doubt that bad genes lead to a frightening society. Segregation and sterilization of these unfortunates is the only way to halt the increase of mental deficiency. Sterilization, Snyder assures us, entails "no discomfort" for men, and for women is not "any more serious than a simple operation for appendicitis." At the end of the chapter, a problem set asks students to outline a practical program of eugenics, to report on their state sterilization laws, and to discuss "from the genetic standpoint" the phrase from the Declaration of Independence, "We hold these truths to be self-evident, that all men are created equal." The staunch advocacy throughout the chapter of eugenic measures, including institutionalization and sterilization, leaves no doubt that for Snyder, the correct answer was that the phrase is biological nonsense.[49]

Allan's, Snyder's, and Macklin's case for introducing genetics into medicine coincided with a general easing of organized medicine's antipathy toward heredity and eugenics. In almost every issue of the American Medical Association's flagship journal during the early 1930s, doctors discussed eugenics questions from immigration to compulsory sterilization, from heredity in gastrointestinal disorders to cancer, from iatrogenic racial degeneration to the improvement of the race. The *Journal* received regular reports from London's Eugenics Society and from eugenic geneticists in Nazi Germany. They discussed the German sterilization law,

Mongolian idiocy (today, Down syndrome), and marriage between relatives. In short, physicians were, by the thirties, actively debating exactly the same range of eugenic questions that the geneticists had in the teens. Their knowledge of genetics was limited, and their therapeutic responses tended to be conservative and old-fashioned—one doctor writing in for advice about recurrence of Mongolian idiocy was told, "It has been the habit of some specialists in such and similar cases to administer throughout the pregnancy the old alternative tonics—mercury, arsenic, iodine, iron and calcium." But evangelists such as Allan, Snyder, and Macklin were able to whip up a sentiment already stirring in the medical community. They hoped to channel it into productive avenues, such as medical education, clinical practice, and research. The process took longer than they hoped, but ultimately this is precisely what happened.[50]

* * *

In the twenties and thirties, then, geneticists' growing disillusionment over the prospects of human hereditary improvement was offset by growing public enthusiasm. Physicians cautiously climbed on the bandwagon, encouraged by proselytizers such as Allan, Snyder, and Macklin. An epidemiological shift from acute, mostly infectious diseases to chronic and hereditary disease, the medicalization of (negative) eugenics, and the wider availability of birth control converged to make "pathological heredity" a topic of widening discussion in American medicine. As a result, negative eugenics, long associated with the elimination—or even, in extreme cases, extermination—of the antisocial or simply foreign elements of society, slowly began to acquire a new connotation. Medical eugenics was preventive genetic medicine, or genetic public health: the prevention of the birth of individuals who would suffer, or cause their parents to suffer. Out of this impulse came the field named medical genetics.

This new biomedical specialty was by no means incompatible with traditional eugenic ends, but it represented a reversal of cause and effect. Francis Galton had sought the amelioration of man through human hereditary improvement. The new breed of medical geneticists sought improvement through the amelioration of man. Physicians had generally opposed eugenics as social control, but they supported it as a means of reducing disease incidence. Subsequent social improvement became a happy consequence of applying medical approaches to hereditary conditions and traits. As long as their primary objective remained the reduction of individual suffering, physicians could become enthusiastic supporters of long-standing eugenic goals such as reducing the burden on society of the unfit.

The invention of medical genetics enabled physicians, medical scientists, public health workers, and others to shift their primary target from the Galtonian population to the Garrodian individual. It was perfectly possible to address human suffering from hereditary causes and treat long-term eugenic goals as a hopeful product of good medical practice. For others, improving the population remained the ultimate goal, with concrete medical practice as the most practical means to an end. Indeed, as Charles Davenport had recognized in 1912, and Irving Fisher a few years later, the individual and the population were but ends of a spectrum. A comprehensive program of medical heredity would encompass the full range. What medical genetics did not do, however, was to drive eugenics out of the study of human heredity. Indeed, eugenics was the means by which genetics went medical.

4
The Heredity Clinics

SOMETIME IN 1937, HARRY LAUGHLIN, the superintendent of the Eugenics Record Office, had a seizure while driving down Main Street in Cold Spring Harbor. He suffered from epilepsy, which had ever been on the eugenicists' list of dysgenic mental conditions. By his own definition, in other words, Laughlin himself was unfit. He survived the accident but was badly shaken. His institute was also a bit wobbly. After Charles Davenport retired in 1934, Albert Blakeslee, the new Cold Spring Harbor director, had launched a review of the ERO, assembling a committee that included Columbia's L. C. Dunn and the Harvard anthropologist Earnest Hooton. They found the record collection to be nearly useless for the study of human genetics, the *Eugenical News* to be an unscientific embarrassment, and the office itself an anachronism. Laughlin's response to this upbraiding included accepting an honorary doctorate from the University of Heidelberg the following year—tantamount to the Nazi seal of approval. In 1938 Laughlin's ally John C. Merriam retired as president of the Carnegie Institution. His replacement, the MIT physicist Vannevar Bush, wasted little time in diagnosing the disease and recommending therapy. The necessary surgery was performed and, on the last day of 1939, the ERO shut its doors.[1]

The demise of the ERO makes a convenient tombstone for the eugenics movement. It and Laughlin have come to symbolize what had gone wrong with Eugenics. Yet in the late thirties Laughlin had begun to push eugenics toward more medical applications—a direction often seen as the antidote to the discredited social, behavioral eugenics of which he himself is the emblem. He sounded out the constitutionalist George Draper about adding a genetics component to his Constitutional Clinic. He petitioned the Rockefeller Foundation and the Carnegie Institution of Washington. And he approached New York philanthropists such as Wickliffe Draper and Frederick Osborn about establishing a medical genetics clinic, perhaps at a medical school such as Johns Hopkins, perhaps adjacent to the Eugenics Record Office at Cold Spring Harbor. But no hospital would join forces with him, and Cold Spring Harbor had no medical accreditation. Laughlin never got his clinic. Nevertheless, he sensed that strengthening the medical aspects of eugenics would have been good reputation hygiene.[2]

He was not the only one. Before the dust had fully settled at the ERO, some of Laughlin's acquaintances created a new, hybrid kind of eugenics institute: the heredity clinic. Heredity clinics were conceived as eugenics record offices attached to medical schools and teaching hospitals. Established on the eve of America's entry into the Second World War, they struggled in their first years. But though they were small in staff and few in number, the heredity clinics represented exactly the sort of integration of eugenics and medicine that Macklin, Allan, and Snyder had been campaigning so hard for since the late 1920s. And small wonder, because they—in particular, Snyder and Allan—were architects of the new hybrid institution. Although their pedigree is unquestionably eugenic, the heredity clinics established the core methods, built the institutional structures, and founded the professional community that we associate with "modern" (noneugenic) medical genetics. In short, they show how interwoven were the twin goals of relief of suffering and human improvement.

* * *

The North American human genetics community in the late 1930s was small indeed. Most trained geneticists had fled human genetics for more tractable experimental organisms such as mice or fruit flies. Yet a few geneticists maintained an interest in humans on the side. The mouse geneticist C. C. Little had been interested in human genetics and eugenics since the teens; he often attempted to relate the mouse work to humans, in particular to cancer. At the University of Michigan, Lee Raymond Dice did behavioral genetics in mice and population studies in humans. At Wisconsin, the zoologist Michael Guyer dabbled in human genetics. In between reporting baroque and brilliant experiments on fruit flies, Hermann Joseph Muller, one of the founders of classical genetics, wrote dire treatises on the accretion of silent lethal mutations in the human gene pool. His bleak, socially conscious genetics rubbed off on his students H. Bentley Glass, Clarence "Pete" Oliver, and, later, Elof Axel Carlson, all of whom went on to write about human genetics and eugenics.[3] Many of these zoologically trained researchers found themselves pulled by their interest in human heredity into the medical sphere.

There were also a few dedicated students of human heredity, mainly with backgrounds in the health and human sciences. Madge Macklin, of course, continued her studies on heredity and disease at the University of Western Ontario. In Charlotte, North Carolina, William Allan was still studying the "defectives" of Appalachia. The biometrician Raymond Pearl, himself increasingly marginalized as a loose cannon, continued his work on eugenics and human biology at the Johns Hopkins School of Public Health and Hygiene, even while savaging the Eugenics movement in speeches and in print. At the University of Chicago, the psychologist H. H. Newman's 1937 book with F. N. Freeman and K. J. Holzinger defined the modern methods of twin research. The Harvard anthropologist Earnest Hooton, though mainly interested in questions of race, regularly consorted with health professionals in conferences and symposia. In 1936 Franz Kallmann, a German émigré, had established a human genetics department at New York Psychiatric Hospital in New York City.[4]

But the most prominent figure in American human genetics in the thirties was Laurence Snyder of Ohio State. Through his work on blood groups, Snyder united the worlds of medical heredity and populational human genetics. Although his contributions to statistical population genetics paled in comparison to those of Wright or J. B. S. Haldane, Snyder was well liked and energetic, and a natural teacher, networker, and administrator. He was passionate about medicalizing human genetics. He became a hub in an emerging network of North American medical genetics. Frederick Osborn, who took the torch of American eugenics from Charles Davenport, called him a "first-class, second-rate man."[5]

Osborn himself was decidedly first-class. Born into a distinguished and wealthy New York family, he had a gentle patrician bearing accentuated by his six-foot-eight-inch frame. He had instant eugenic credentials: he was a nephew of Henry Fairfield Osborn, a founder of the Galton Society, the elite anthropological eugenics club that still convened at the American Museum of Natural History in New York. Henry Fairfield Osborn's son Fairfield Osborn was a conservationist and the head of the New York Zoological Society. The cousins Fairfield and Frederick collaborated on population problems, although Frederick was also interested in medicine. In 1928 Frederick Osborn retired from business and began to soak himself in eugenics, learning its philosophy at his uncle's knee. He joined and soon became an officer in all of the New York–area eugenics societies: the Galton Society, the Eugenics Research Association, the American Eugenics Society—he was the leading voice in the AES until the 1970s. He was on the board of the Carnegie Corporation and the Carnegie Institution of Washington, and he was part of the network of New York finance and philanthropy. Using these contacts, Osborn set about reorganizing American human genetics and eugenics. He sought to ensure eugenic progress, curb population growth, and promote biomedical research—while keeping his own reputation clean of unsavory associations. Where Snyder was the social hub that connected a disparate and far-flung group of workers interested in human heredity, Osborn was an idea and a money man. He reconceived American eugenics, adroitly reading the increasingly medical direction it was taking, and became the conduit through which the money would flow to support it. Working both independently and together, Snyder and Osborn established the first heredity clinics, reconfigured American human genetics, and established a newly medical eugenic agenda.

Morale was low among eugenics activists in the early thirties. Despite increasing medical interest in the subject, membership in the American Eugenics Society declined through the first half of the decade. Osborn later noted that much of this attrition occurred among the ranks of prominent and wealthy citizens who had contributed much to the vigor, visibility, and zeal of the society in the 1920s—and who had done much to cheapen its reputation among scientists. No doubt the Depression was a major cause of the decay. But the scientific membership fell away as well. Only seventy-three people had attended the Third International Congress of Eugenics in New York in 1932.[6]

Yet by the time Osborn took the reins, eugenics was experiencing something of a revival. The cultural *terroir* was becoming more suitable for eugenically minded programs. Since *Buck v. Bell*, many American states had passed new sterilization laws designed to withstand constitutional challenge. The number of sterilizations performed increased steadily through the decade, more than doubling from 16,066 in 1932 to 33,035 in 1939. In Michigan in 1935, the state Academy of Arts, Science, and Letters called for the creation of a Family Research Bureau, to "study the causes of criminality and of mental defects and diseases, and if possible discover effective means for preventing and curing such conditions." Support for birth control also grew steadily. In 1936 Margaret Sanger successfully challenged the classification of birth control materials as "obscene" under the Comstock Laws; the next year, North Carolina passed the nation's first law explicitly permitting state agencies to distribute contraceptives and contraceptive information. The motivation behind the push for relaxed birth control regulation included factors having nothing to do with eugenics, particularly women's demand for the right to regulate their own reproduction and family size. Through such measures, sterilization and birth control became increasingly entwined; by the 1960s voluntary sterilization would become one of the most popular methods of birth control. Such changes made the cultural soil more hospitable to the eugenic seed.[7]

Also in 1936, in New York the wealthy financier Wickliffe Draper consulted with Laughlin, Osborn, and others in creating a philanthropic structure for the support of eugenic causes of all sorts, but especially those related to race and preventing miscegenation. One component of this fund was to have been a "marriage clinic," from which people could seek advice—mainly about race and disease—in regard to potential mates and to childbirth, although this quasi-medical aspect was eventually dropped as the group crystallized as the Pioneer Fund. In Minnesota in 1937, the eccentric physician Charles Fremont Dight died, leaving a modest endowment to the University of Minnesota for the creation of a new eugenics institute. California's environmentally oriented, family-based eugenics network was thriving, with Paul Popenoe's marriage clinic in Los Angeles; Samuel J.

Holmes up in Berkeley and highly active in the AES; and the eccentric, zealous, and pro-Nazi retired banker C. M. Goethe in Sacramento financing eugenics efforts in myriad ways. In the late thirties, then, there was a widespread sense of optimism and rebirth among the movement's leaders, with ambitious plans and money available to fund them. By the time the writing was on the wall for the Cold Spring Harbor Eugenics Record Office, the eugenics leaders could take its decline as an opportunity for restructuring.[8]

* * *

Through the quirks of personal connections, the first focus of that restructuring was central North Carolina. Both Osborn and the North Carolina physician-eugenicist William Allan were on the board of the Eugenics Research Association, the research and publicity arm of the Eugenics Record Office. When the AES was created in the 1920s, Irving Fisher and Charles Davenport had defined it as primarily an educational and propaganda institution, with the ERA serving as the movement's research arm. The ERA held an annual meeting and published *Eugenical News,* ten or so scattershot pages chocked with research updates on the biochemistry of sperm or the material nature of the gene, clinical and public health studies on incidence of disease, legal news on marriage legislation, and utopian mini-rants about our eugenic future. When it became clear that the ERO would be shut down, Osborn seized the opening as an opportunity to realign the axes of American eugenics.

Following a 1936 ERA meeting in New York, Allan approached Osborn with a scheme: "What is in my mind is this; someone ought to start a eugenics office as part of a county health department somewhere, and this seems a good place to do it." It was a genteel southerner's plea for financial help. As a physician in private practice, Allan had been unsuccessful at obtaining support from either the foundations or the public health agencies. He suggested that Osborn relocate from Manhattan to Charlotte to set up the clinic—a naïve idea, if he intended it literally, but canny if this was the desired effect: Osborn demurred on moving to North Carolina, of course, but said he would very much like to visit and receive a eugenic tour of the region. That visit, followed by many further conversations and much correspondence, resulted in Osborn's brokering a series of agreements among

Allan, the Carnegie Corporation, and the new Bowman Gray School of Medicine of Wake Forest University, which resulted first in substantial funding of Allan's eugenics research and later in attaching Allan and his small team to an academic medical school, leading to the founding of the first department of medical genetics in the country. At the end of his career, Osborn proudly listed the Wake Forest program as one of the fruits of his efforts and as an indicator of the new direction in eugenics that he had spearheaded.[9]

Stimulated by Allan's perspective, Osborn reconceived the division of labor within the movement. The "old" eugenics had moved along the common axes of the Progressive era—race, ethnicity, and social class. Now, at the tail end of the Great Depression, Osborn and Allan discussed realigning those axes. "The groove I move in is a very narrow one," Allan wrote to Osborn in 1937, "in which I constantly encounter those who have inherited physical or mental disaster (blind, deaf, crippled, etc.), and these derelicts constitute such a serious problem that I perhaps over emphasize the genetic end and think too little of the lack of good environment." Osborn took the point. "I am in an area of declining births, where the loss of good genes has us all worried. You are in a high birthrate area, where the constant increase of bad genes is a cause not only of anxiety but of daily observation and distress."[10]

In short, Osborn and Allan saw positive eugenics as urban and social, negative eugenics as rural and medical. From his New York offices, Osborn himself planned to lead the AES in a mixture of eugenics education and social programs in welfare and public health designed to increase the frequencies of the "best" genes in all sectors of society. Given recent excesses in the name of negative eugenics, it is understandable that eugenic enhancement seemed gentle, social, friendly. The other, less politically palatable side of eugenics—elimination of the unfit—would be handed over to trusted, soothing doctors. We have seen that negative eugenics had been going medical for several years. Osborn's genius was recognizing that trend, encouraging it, and pairing it with a seemingly benign new approach to positive eugenics, or enhancement. "We don't for a moment mean to imply a neglect of negative eugenics," Osborn insisted. But negative eugenics needed a friendly, trusted face. Osborn understood that physicians had both the tools and the moral capital to carry out a program of negative eugenics. As he wrote in his 1940 book *Preface to Eugenics*, negative eugenics was "the concern of the medical profession and public health authorities. Their complexity requires that they be in professional hands, and it is unlikely that the public will accept advice from any other source." The new negative eugenics, then, would focus on health rather than sociality. Disease would now be the prime target, although that included "feeblemindedness" and other mental conditions on the traditional eugenics agenda. The umbrella of health has always been capacious and flexible; few old eugenic targets needed to be dropped, although some had to be renamed.[11]

In Osborn's scheme, the husk of the Eugenics Research Association was the perfect receptacle for a new, medically oriented organization for negative eugenics. The first step in reorganizing the ERA was the election of Laurence Snyder to the presidency in 1937. He succeeded C. M. Goethe, who epitomized the old eugenic guard. Snyder was the perfect transitional figure: his Harvard Ph.D. in genetics gave him scientific credibility; he considered Davenport one of his primary mentors; and he was a pioneer in the introduction of genetics into medical schools. The next year, Osborn wrote to Allan about reorganizing the ERA along medical lines. It would need a new name, something bland like the Association for Research in Hereditary Factors, perhaps, "because the word 'eugenics' has so many connotations which are confusing to many medical men." Notwithstanding that "confusion"—clearly a euphemism for misguided opposition—over the word, Osborn was convinced that doctors were natural eugenicists. Echoing Davenport in 1912, he told Allan that negative eugenics "should be increasingly a science in the hands of the medical profession for application." With negative eugenics under the impeccable stewardship of America's trusted physicians, Osborn was free to return to Galton's original concept of a primarily positive eugenics driven by education and incentive, with coercion necessary only in extreme cases of incapacitation or pigheadedness.[12]

Allan could not have agreed more; he had written as much in recent articles such as "Medicine's Need of Eugenics" (1936) and "The Relationship of Eugenics to Public Health" (1937). The conquest of infectious disease, he wrote, required bacteriologists who would eliminate pathogens from water and milk supplies, immunize the populace, quarantine sick individuals, and control animal and insect vectors. In contrast, to discover and control hereditary disease was simple: it required "nothing more elaborate than a family record office in the county health department." Osborn gave Allan that family record office.[13]

In 1938 the Carnegie Corporation (on whose board Osborn sat as well) initiated a new funding program. Allan leapt at the chance. "A $3,000 grant-in-aid," he told Osborn, "would give me a chance to put out a field worker to survey this territory in a systematic manner, instead of depending on my finding odd moments to dash off and bring home samples." Since Davenport's day, fieldworkers had been unskilled but trainable researchers, almost all women, who collected but did not analyze data. Allan got more than he asked for. Osborn introduced him to Charles Dollard, assistant to the president of the Carnegie Corporation. The Carnegie would provide the funds, but Dollard and Osborn thought Allan needed the collaboration of a fully trained geneticist. This move was probably a tacit acknowledgment of Allan's limited genetic training—something Allan admitted, though he tended to portray it as a strength. Further, the plan would bring new men into human genetics. Allan was a good mentor—rugged, warm, appealing—and a practical geneticist, a eugenicist, full of craft knowledge. Osborn understood both Allan's merits and his deficiencies as a medical-genetic pioneer.[14]

Osborn and Dollard often consulted Leslie C. Dunn, a highly respected *Drosophila* geneticist uptown at Columbia University, on genetic matters. They hoped Dunn would be able to suggest someone with knowledge and reputation strong enough to withstand scrutiny from skeptics. Allan was agreeable to the plan, though he cautioned that for fieldwork, temperament was as important as technical knowledge. "If you and Professor Dunn picked a man with your gift of friendliness it would be smooth sailing," he told Osborn. "But a man with Laughlin's shyness and inhibitions, for instance, would be mistaken for one of General Sherman's stragglers and strung up." Allan naturally suggested his old collaborator Laurence Snyder as his genetics adviser, citing Davenport's opinion that Snyder was the leading human-geneticist in the country. Osborn delicately yet firmly rejected the nomination. He himself liked Snyder, he said. However, "the opinion of the geneticists themselves," unfortunately, "does not apparently place Snyder in the top flight in which such men as Sewall Wright of Chicago and L. C. Dunn of Columbia are leaders. It is my impression that the geneticists think of Snyder a little the way they do about Davenport, though to a much less degree, and are inclined to think of work on human heredity as being the kind of work done by men like Davenport and Snyder, with some rather indefinable derogatory implication." He continued, diplomatically, "Now, I think that's a rotten attitude on their part, but I gather there is just enough justification for it to make it hard to combat directly." A renewed effort in medical genetics and negative eugenics needed to be above reproach scientifically. Dunn suggested his former student Paul R. David, a geneticist then at Lafayette College in Pennsylvania's Lehigh Valley. Although Allan admitted that "the name startled me a little"—Jews? In North Carolina?—Dollard reassured him that David was from "these parts" and was a "first class fellow." Allan agreed to take him on.[15]

North Carolina was an ideal test ground for medicalized negative eugenics. The state had long been in the vanguard of hereditary and sexual public health policy. It had had a sterilization law on the books since 1929. And when the Supreme Court struck down the federal proscription against distributing contraceptives in 1936, North Carolina was the first state to pass a law specifically allowing it. Immediately, the soap powder heir Clarence J. Gamble established a program to disseminate condoms and his contraceptive foaming powder to rural North Carolinians.[16] Osborn was already interested in population problems at this time and doubtless had been following these events with interest. Probably no state in the Union had conditions more favorable to his vision of a medically mediated negative eugenics.

Allan considered his eugenics work akin to that of a hereditary public health officer. With Carnegie support, in 1939 he opened the Family Record Office in Charlotte. He soon formed an advisory board for the FRO that included the state public health officer and the deans of medicine at the three North Carolina medical schools: Duke, the University of North Carolina, and Wake Forest. In March 1940 Allan visited all of them, to feel them out on the possibility of getting an academic program in medical genetics started.[17]

He found Dean Coy C. Carpenter at Wake Forest the most receptive. Wake Forest medicine was in a state of fertile upheaval. Carpenter was overseeing the conversion of its two-year medical school into a full four-year program; simultaneously, the medical school was moving across the state, from the Raleigh area to Winston-Salem, thanks to a bequest from the tobacco magnate Bowman Gray, head of R. J. Reynolds. The bequest funded only the establishment of the new school—it provided no endowment. Carpenter had a lot of freedom but no money. When Allan discussed implementing a novel program in medical genetics with outside funding, Carpenter listened eagerly.[18]

Allan was sixty years old and his health was weak. If Carpenter was going to get a permanent program in medical genetics, he would need a successor for Allan. He suggested that if Allan could obtain additional Carnegie funding to support a bright young resident to serve as second in command, he would create a new department of medical genetics when the new school opened in Winston-Salem. Allan would gain a university appointment, a protégé, and a genetic consultant; Carpenter would have a progressive, promising, and best of all completely funded new program in medical genetics to help his new school make a splash; and Dollard and the Carnegie would be bringing new men into human genetics while supporting a quiet program in eugenics in a medical school, where it would be infused with the sanitary odor of carbolic acid.

Claude Nash Herndon, Carpenter's pick for Allan's junior associate, was twenty-two, born and bred in North Carolina, with a bachelor's from Duke and an M.D. from Jefferson Medical College in Philadelphia. He had a genial southern manner, polite and deferential, but he had an earthy side that came out in his private correspondence. In September 1940 Carpenter sent him to spend a week in the field with Allan to see how he liked the work. "After four days of pretty strenuous field trips," Allan told Carpenter afterward, "Dr. Herndon tells me he likes the outlook in genetics and wants to tackle the job." Herndon went on to be one of the leaders of the "new" human genetics of the nineteen fifties and sixties.[19]

Research was necessary but not sufficient; any comprehensive program in human heredity also had to reach students. "All the deans" of the North Carolina medical schools, Allan wrote to Osborn, "want some genetics taught in their schools." Chances are it was Allan who convinced them. Instructors in medical genetics were hard to find, however. Allan knew of just one: Snyder. Whatever Snyder's reputation as a scientist, he was unquestionably the most qualified instructor of medical genetics in the country. Since 1933 he had been teaching the nation's only required course in genetics in a medical school. Allan invited him to give it, in triplicate, to North Carolina's medical students. In 1941, on Carnegie funding, he spent a mild southern winter commuting between Raleigh-Durham and Wake Forest, offering a series of lectures based on his Ohio State course. Duke University Press published the results almost immediately.[20]

In between lectures, Snyder consulted with Allan, Osborn, Dollard, and Carpenter as they hammered out the details of the new program in medical genetics at Bowman Gray. It would be called the Department of Medical Genetics, although administratively it would reside within the larger Department of Medicine—thus it was more like a division in many medical schools. It would open in the fall of 1941, with the new medical school. Allan and Herndon would continue their research. The hospital affiliation would provide a reputable outlet for clinical services and eugenic advice. And the medical school would provide a forum for teaching genetics to medical students. The program was designed from the bottom up as an integration of genetics into Allan's clinical research program of eliminating serious hereditary disease from families. Consultation with the geneticist Paul David would inject needed genetic theory into their work.

The only weak spot in the plan was Herndon's lack of training in formal genetics. At first, it was thought he might spend a year or so with L. C. Dunn at Columbia, but then a fortuitous new opportunity arose. In July 1941 Allan received a letter from Charles Cotterman, a student of Snyder's now in Ann Arbor, Michigan, where he was working in the Laboratory of Vertebrate Biology under director Lee R. Dice. Cotterman had written to ask whether Allan had a medical man with an interest in genetics who might be looking for a job and some training. It must have seemed providential.

* * *

Lee Raymond Dice came to medical genetics from a background in zoology and the West Coast conservationist style of eugenics. Born in Georgia in 1877 (he was, he believed, named after Robert E. Lee), Dice was raised in rural Washington State. He went to Stanford, where he studied evolution with David Starr Jordan and Vernon Kellogg. He then went to Berkeley for graduate school, where he worked with the eugenicist and ecologist Samuel J. Holmes. He served as a deputy fur warden for the Alaska Fisheries Service, did fieldwork in eastern Montana, and caught rats in Washington, D.C., for the U.S. Biological Survey. Following military service in the First World War and brief stints at several universities and colleges, in 1919 Dice was offered a position in zoology and the curatorship of mammals in the Museum of Zoology at the University of Michigan in Ann Arbor. He never left. In 1925 C. C. Little became president of the university. He soon established a Laboratory of Mammalian Genetics and put Dice in charge. By the thirties Dice had begun a research program in eugenics and human genetics. In 1937 he responded to a call put out by the state's Academy of Arts, Sciences, and Letters for Michigan to create its own Eugenics Record Office. Michigan already had a tradition of eugenics and human genetics—Battle Creek was an hour to the northwest, and Victor C. Vaughan made his home in Ann Arbor. It was a natural fit.[21]

Dice needed staff for this new effort, and he himself was not trained as a geneticist. Laurence Snyder again provided the link. His student Charles Cotterman was among the brightest, most original, and most eccentric individuals in genetics, arguably "the last universal mind in biology," in the words of his obituarist and former colleague John Opitz.[22] For all his brilliance, though, Cotterman had a stormy relationship with publishing. His dissertation, "A Calculus for Statistico-Genetics," has been called the "most famous unpublished paper in the history of human genetics." He had hoped to finish in 1939, but in late 1940 he was still working on it, even as he was getting settled into his first faculty position.[23]

In late fall 1940 Dice formulated a plan for a "department of human heredity" that would expand his and Cotterman's work in human heredity and operate in "close association with the Medical School," including privileges in the University Hospital. The problem was, both Dice and Cotterman were Ph.D.s and thus could not get access to the medical records they needed for data. Neither had an interest in any particular medical conditions; what they wanted was reliable data on human traits. Dice had connections in medical departments: Harold Falls in ophthalmology and George Moore and Byron Hughes in the orthodontics clinic. But his proposed

4.1 Lee Raymond Dice, founder of the Michigan Heredity Clinic. Courtesy of Bentley Library, University of Michigan

heredity clinic needed a medical man on staff. Beyond that, Dice was developing a commitment to medical genetics as a hybrid of science and medicine that required formal training in both fields. Over the coming years, he sponsored or encouraged a great deal of cross-pollination between scientific genetics and academic medicine, one of the key themes of medical genetics.[24]

By sending Herndon to Michigan, Allan could gain the needed training in genetics for Wake Forest and simultaneously expand the nascent field by helping to set up a second program in medical genetics. F. P. Keppel, at the

Carnegie Corporation, was willing to release Herndon to go to Michigan. "We are interested in what you are doing at Michigan," he assured Dice, "and if releasing Herndon will help the cause which we are all in one way or another promoting, the Corporation is glad to cooperate." Herndon packed his bags and arrived in Ann Arbor at the end of August.[25]

In the fall of 1941, as it was becoming increasingly obvious that the United States would not be able to stay out of the war, American human genetics was similarly in a state of ferment. The Bowman Gray School of Medicine opened in September, just as Herndon left for Ann Arbor. When he arrived, he found himself near the heart of a burgeoning center of North American activity in human genetics. Snyder was just a few hours away in Columbus. And, as Herndon informed Allan, the University of Minnesota had just formally opened the Dight Institute of Human Genetics. Like the Michigan Heredity Clinic, the Dight Institute operated out of the Department of Zoology but maintained close contact with the medical school. It was directed by Clarence "Pete" Oliver, *Drosophila* geneticist and student of H. J. Muller from Texas. The three institutes, born as nonidentical triplets, shared both the bonds and the rivalry of any siblings. The Dight, however, was the odd brother out. Whether it was Pete Oliver's personality or simply the lack of personal connections with men like Snyder and Osborn, the Dight was never as close to either the Bowman Gray department or the Michigan Heredity Clinic as the latter two were to each other. All three, however, shared the sense that hybridization with medicine was the route to success. "Oliver told Cotterman that one of his main objects was to arouse interest in genetics among the medical people," Herndon wrote Allan in September. As we shall see, their visions of what the fusion of genetics and medicine ought to look like often clashed. The making of the first medical genetic institutions was the result of the efforts of men whose ideologies were rooted in clinical medicine, theoretical genetics, and eugenics, respectively, negotiating with one another to form a new, medically oriented kind of Eugenics Record Office.[26]

* * *

The clinically trained Herndon was unimpressed by academic genetics in Ann Arbor. Cotterman had a Galtonian turn of mind—statistical, populational. He could not have cared less which traits he studied, so long as they could be analyzed clearly. Medical patients were simply a good source of material for genetic analysis. Herndon, in contrast, wanted to treat real diseases and prevent suffering. At the end of his first month in Michigan, he wrote to Allan, "I certainly wish I could get these folks to listen to some real practical genetics and get them interested in things that are really serious. They are about to drive me dippy with such things as anomalies of the eyebrows and minor defects of the teeth." Allan chuckled at Herndon's letter and replied, "We'll ruin the academic outlook someday I hope." Tensions notwithstanding, Herndon got to work. By the end of November he had helped Dice and Cotterman get their clinic open, and he

4.2 Claude Nash Herndon, holder of the chair in Medical Genetics at Wake Forest School of Medicine, endowed by the eugenic philanthropist Wickliffe Draper. Courtesy of Wake Forest School of Medicine

sent Allan an announcement flyer, adding sourly at the bottom a paraphrase of Sextus Empiricus's *Against Professors*: "The mills of the Gaels grind slowly. It has taken me this long to get this far."[27]

Two weeks later, the Japanese struck Pearl Harbor. Before Christmas, American medical schools were instructed to shift to an accelerated, four-quarters-per-year program, to increase the flow of much-needed physicians into the military. The three new programs in human genetics—and the entire fledgling enterprise of the Bowman Gray School of Medicine—were destabilized as soon as they began.

One almost immediate effect of the war was to spring Herndon from Michigan. Deeply dissatisfied, over the Christmas holiday he apparently had a heart-to-heart with Allan about returning to Winston-Salem. Herndon wrote him a short note on New Year's Eve, saying, "Dear Dr. Allan, You win. I'm going back to Michigan. But I am still of the opinion that if that is genetics I would rather be practicing medicine in the backwoods." The following February, however, Coy Carpenter did a bait and switch. He first obtained Herndon's release from Michigan on grounds of military service, then secured a deferral on the grounds that he was essential teaching staff at Bowman Gray. In the result, Herndon's "permanent" position in Ann Arbor lasted just six months. By April 1942 he was back in Winston-Salem, teaching in the medical school and doing the fieldwork that he loved. Allan and Herndon were tickled.[28]

The two units tried the reciprocal exchange as well. Dice lost both Harold Falls and Charlie Cotterman to the war effort, but in late 1943 Cotterman secured a transfer to Bowman Gray Medical School, where he enrolled as a first-year medical student in January 1944. Cotterman fared no better in medical school than Herndon had in the laboratory. Writing from Winston-Salem to the Heredity Clinic secretary Nedra Kuntz, he captured in two words the experimentalist's boredom with the medical curriculum: "Right now (7:00) I must be running along to a 'Clinical Pathological Conference.' Ho hum." He lasted only a year, saved in 1945 by another transfer and then the end of the war. The hybridization of science and medicine was fraught with tensions of personality, culture, and intellectual style.[29]

At Ann Arbor, Dice could not hold on to his staff and could not replace them: the military sucked the universities almost dry. And he badly needed more funding to support the various research efforts of his Laboratory of Vertebrate Biology: the Heredity Clinic, his behavioral genetics project with the dormouse *Peromyscus*, and various ecological projects. He drummed up business with promotional articles about the Heredity Clinic in local papers—"U. of M. Heredity Clinic Seeks to Improve Human Breed"—and even *McCall's* magazine. He also sought a replacement for Herndon. He wanted badly to find someone with both an M.D. and a Ph.D. in genetics. One possibility was Ray Anderson, a student of Pete Oliver's who came to Ann Arbor to do an internship in pediatrics. He cooperated with the Heredity Clinic, but then he too was called into service, and when he came back to the States he returned to Minneapolis to work at the Dight. The Dight Institute itself, with independent funding and a staff of only two, did not have a high enough profile to be greatly affected, although it too weathered a crisis just after the war. In 1946 Pete Oliver left abruptly for a post at the University of Texas, Austin, rendering the Dight dormant for more than a year while it searched for a new director. Sheldon C. Reed, a geneticist who, like Snyder, had trained under William C. Castle at Harvard, took over in 1947 and ran the Institute until 1975. Reed's views on eugenics, medical genetics, and genetic counseling were complex. He is considered the founder of the field of genetic counseling; he coined the term in 1947, in order to distinguish it from previous eugenic practice. Yet he saw heredity clinics such as the Dight as the most effective way to bring about eugenic improvement. In short, Reed, like his compatriots, was an unabashed but complex eugenicist who saw medicalization and patient autonomy as critical for the fulfillment of eugenics' promise.[30]

Finally, one must note that during the war, Charles Davenport died quietly, on February 18, 1944, a victim of environment and temperament. The old marine biologist was an avid supporter of the local whaling museum. Overworking himself while cleaning a rare beached whale, he caught and succumbed to pneumonia. The official announcement from Cold Spring Harbor made no mention of eugenics or the ERO, even though just two years earlier, Laboratory director Milislav Demerec had tried to reopen it, even palpating H. J. Muller and a young James Neel for the post. The ERO was high and dry.[31]

* * *

The groups at Ann Arbor and Winston-Salem managed to eke out some research during the war years. Allan and Herndon continued their field studies, funded at first almost entirely by the Carnegie Corporation. Their main project was a survey of hereditary disease in the counties of the Great Smoky Mountains to the west, particularly Watauga County. These mountain counties were a "gold mine in heredity," Allan wrote. It was homogeneous, "a country without negroes, tenant farmers, or shifting mill population." It was reproductively isolated, with little immigration in or out since about 1810. "Practically everybody," Allan claimed, descended from the post-Revolutionary pioneers and so could trace their lineage back four to six generations. They often found relatively complete genealogies recorded in the pages of family Bibles. And there were, he wrote, "unbelievably high rates of in-breeding." This meant that many rare recessive diseases appeared at relatively enormous frequencies, compared with the population at large, and further, "that the carriers of recessive or hidden pathological traits can be traced back to the pioneers who brought these traits into that region, hence the possible carriers of such traits will be confined to the direct descendants of such pioneers." He told Snyder, "It seems to me the greatest natural experiment in inbreeding since the Children of Israel crossed the Red Sea and had to marry their cousins for the next forty years." Twenty years later, the Johns Hopkins medical geneticist Victor McKusick would find an even greater natural experiment in inbreeding among the Amish.[32]

Allan, Herndon, and Florence Dudley—Allan's secretary, whom he trained into an effective and enthusiastic fieldworker—surveyed these populations for a variety of hereditary conditions. Allan's pet diseases were deafness and blindness—including several forms of cataracts and retinitis pigmentosa, a reliable Mendelian recessive—as well as such crippling diseases as muscular dystrophy and peroneal atrophy. He made a study of albinism and in fact was happy to track down almost any disease on which they could get a strong lead. Allan's dream project was begun near the end of his life and was carried on by Herndon after Allan's death. It was a survey of the entirety of Watauga County for every hereditary disease they could document. He had been planning such a study for years, but in December 1941 he proposed it concretely to Osborn, and the Carnegie agreed to fund it.[33]

It was shoe-leather genetics. Allan, Dudley, or Herndon would typically go out into the field for one to several weeks at a time, setting up a base at a local hotel. Allan's favorite for the Watauga Survey was the Daniel Boone Hotel, in Boone, North Carolina. From there, they would make forays into local communities. Often they would contact a local doctor or clergyman and inquire about hereditary diseases. The people could be harder to find than the records. Bad leads, bad roads, bad weather, and wartime shortages all hampered their efforts. "Things have been moving slowly here since my last report, due to several causes," Herndon wrote at the end of a long day. "First, the bottom fell out of the clouds around here this past week, and this county has been practically under water. My other difficulties are poor diagnoses and worse addresses." Gasoline rationing forced very careful itineraries, which were often foiled by an itinerant study population. "So far my luck on this trip has been pretty poor," Herndon reported. "I was in Durham Saturday, and spent most of the day trying to trace a colored girl that seemed to think it was cheaper to move than to pay rent. Finally found her though." Despite Allan's cheerful description, Herndon seemed to think the Negroes that were there pretty shiftless.[34]

Once Allan, Herndon, or Dudley found a subject, they administered a medical examination and took a family history interview. They recorded all data on forms obtained from Cold Spring Harbor. Along with trait and pedigree information, they collected folk theories of heredity. "I think I more than met my match this morning," Herndon wrote.

> I went to see two sisters who are albinos, and listened to a two hour lecture on astrology and got about 15 minutes-worth of pedigree. As soon as they found out what I was up to I couldn't get a word in edgewise. It seems they know exactly what causes albinism and all other kinds of hereditary diseases, and they proceeded to tell me. I gathered that albinos occur when Saturn is exactly 8 degrees from something or other under the signs of Pisces and something else. . . . They are certain that we are wasting our time gathering pedigrees when all we need is a good book on astrology.[35]

Jabbing at their geneticist-consultant, Paul David, he added, "It all sounds rather complicated to me, so I recommend you send Dr. David up to talk to them." In Herndon's mind, his and Allan's style of "real practical genetics" was rigorous and tough-minded; Paul David's theoretical population genetics was only a step away from fortune-telling.

Practical genetics had a natural-history quality. Allan, Herndon, and Dudley collected diseases the way an ornithologist collects birds. "On the Wildcat Road of the Stony Fork this afternoon, I found a nice trait of deaf-mutism," Dudley wrote from the field. Once they had a good trait, they worked out the pedigrees, identifying the affected and the carriers and trying to pinpoint the "pioneer" who introduced the trait. "The retinitis [pigmentosa] pedigree is exactly what we want," Allan wrote in 1942, "as I think we should be able to tag everybody since the pioneer with the correct probability of carrying the trait." Allan attended family reunions, interviewing as many of the attendees as would talk to him. Winning their trust was critical, and Allan drew upon his considerable charm and social skills, gathering jokes and stories he knew people would enjoy. He told Herndon, "I have accumulated a couple of church stories to use on the Baptists." Sometimes he drew the pedigrees on-site and in real time, taping large rolls of butcher paper around the outside of a church and filling in the chart as people told him their stories. With the pedigree method, medicine merged into anthropology: beyond simply identifying the hereditary pattern, Allan and Herndon sought to identify the original person in whom a condition occurred. Allan was familiar with standard population-genetic methods for identifying carriers of a trait (heterozygotes), but he needed to know which individuals were carriers. "The technique of determining the number of heterozygotes by taking the square root of the number in the sample studied does not identify anybody," he wrote to Snyder in 1942. "So we have just taken these rare recessive traits, like gyrate atrophy of the choroid, and chased them back to the pioneer that brought them in here." The direct descendants of that pioneer were possible heterozygotes; analyzing the phenotypes of their offspring could reveal their genotype. "And these folks are all known to us by name, age, sex and location, and the chances of their heterozygosity are usually not hard to determine." Population genetics would have given Allan an accurate estimate of the carriers in a population, but no indication of whether a given individual was a carrier or not.[36]

Knowing the individual carriers, he said, enabled them to give eugenic advice. "This gives us a small number that may need birth control information, should they have already married their cousins. This frees the rest of the county from any fear of carrying the trait." This was practical genetics, what Allan called simply the family history method. All the geneticists' fancy statistics did not tell you why *this* person has *this* disease. His was a down-home, high-touch genetics: "I judge this problem calls for a much more intimate acquaintance with our population rather than any further refinement of methodology," he complained once to Paul David. He was more diplomatic to his friend Laurence Snyder, admitting, "As far as you real geneticists being too theoretical goes, there wouldn't be any science of genetics today if you fellows had not worked your gray matter pretty hard." However, theory needed to be worn lightly, in his view. "Every now and then the suit you men cut out doesn't exactly fit everywhere," he continued, "and I tinker a little on some of the tight spots."[37]

Allan and Herndon also attempted to find linkage, to correlate genetic diseases with known human genes, of which there were fewer than a dozen in the 1940s. If someone had x, what was they chance he would also develop y? What genes were inherited together? One of their best tools was Snyder's PTC test, which they always tried to administer when taking a history. Sometimes, getting people to take the test proved challenging. The rumor spread that the bitter strips had a different purpose. "Word got around in the back country that they were sneaking around and sterilizing" people with the PTC strips, reminisced Allan's daughter, Elizabeth Berger, in a taped conversation with Herndon. In her analysis of the North Carolina Eugenics Board, Johanna Schoen found that some poor rural North Carolina women actively sought sterilization. With the public health nurse Lena Hillard traversing Watauga County at the same time, distributing contraceptive sponges and foaming powder, it is easy to understand how the people might have thought the PTC strip a new birth-control technique—or even a means of sterilization. "And the people didn't want to take it," Berger recalled. "At least some of them didn't." Herndon agreed, adding that Allan "suddenly noticed that, when he'd go to a farm house, that the ladies were always willing to take it and the men lit out for the barns."[38]

Allan and Herndon were not sterilizing people out in the countryside, but the hillfolk were not wrong about the researchers' larger intent. The purpose of the fieldwork was simultaneously medical and eugenic. In a letter to Paul David, sent from the field in 1942, Allan described these twin aims. "What I'm trying to do in Watauga," he wrote, "consists of an effort to solve the riddle of the frequency and distribution of rare recessive pathological genes." But the next sentence described his "primary purpose": "cutting down the supply of defective children." Just as important as the men lighting out for the barns, however, is the observation of the women lining up to take what they think is a new form of birth control. While some women were clearly coerced or tricked into receiving contraception or sterilization, others took birth control eagerly. Allan, and to an even greater extent Herndon, operated in a world without sharp moral boundaries. Providing a welcome service graded insensibly into providing a needed service, which graded into providing education, into providing advice, into persuasion, into coercion. When is intransigence an expression of autonomy, and when a symptom of a desperate need for medical care? Public health eugenics involved a complex physio-moral calculus, which Herndon executed paternalistically and often chauvinistically, but with humanitarian intent.[39]

A fourth project, in addition to surveys of crippling diseases and blindness and the Watauga County survey, was billed explicitly as a "eugenics project" in Forsyth County, of which Winston-Salem is the county seat. Reporting on the year's activity to the Carnegie Corporation for the year 1943–44, Herndon wrote that the project, begun in September 1943, was aimed at "the eugenic improvement of the population of Forsyth County." It consisted of a "gradual, but systematic effort to eliminate certain genetically unfit strains from the local population." A principal technique, of course, was sterilization. About thirty sterilizations had been performed, Herndon wrote. Eugenic sterilization in North Carolina was treated as an aspect of public health. The reasons for sterilization were many, and the impetus might come from doctors, the state, or the woman in question herself. "Feeble-mindedness" continued to be a common official reason for sterilization. The North Carolina Eugenics Board records document horrific examples of women and even girls as young as ten being "diagnosed" as feebleminded and sterilized against their or their family's wishes. Allan himself appears not to have thought much of feeblemindedness as a genetic trait. In one of his chatty field reports to Herndon in 1942, he wrote, "The feeble-minded child I will try to get a look at, but I am a little shy about this trait as I don't know how much of a factor heredity is." He preferred traits with clearer hereditary patterns. "Good old retinitis pigmentosa," he continued, "we can go to town on, since it is 100 percent hereditary." This was exactly the kind of solid footing on which Frederick Osborn had been hoping to place negative eugenics. Allan felt confident exercising his authority as a doctor to eliminate unfit strains of Tarheel humanity, but he tried to be careful that the traits he selected against were both medically and genetically well defined. The germ theory of genes gave better medicine.[40]

Allan died, suddenly but not unexpectedly, on April 24, 1943. Herndon began working in the hospital's outpatient department, both as a form of service and for much-needed income—for several years this paid his entire salary. It left little time or energy for research, and Herndon did much less fieldwork. Increasingly, he drew on "clinical material"—patients—he encountered in the outpatient department and through referrals for his research. When he found a new and interesting case, he would take the patient's pedigree and then often go out into the field to try to track down family members, who might be examined at home or, if it was convenient, brought in to the clinic and examined there. This was very much the style of the medical genetics clinics that sprung up in the 1950s and 1960s. Thus Herndon's shift from old-style eugenics surveys to clinical genetics was quick and smooth, a simple decision prompted by a death and a war.

The final dimension of the Wake Forest team's work was education and propaganda. Allan gave numerous talks and lectures to public groups. He and Herndon gave "talks on negative eugenics" at civic clubs, medical societies, and other organizations. Herndon and Florence Dudley brought poster exhibits to medical and public health meetings and made it a point to network with anyone likely to be sympathetic to their positions on birth control and sterilization. Herndon would talk to *almost* anyone. From a conference in St. Louis, he wrote home to Florence Dudley, "Three Catholic sisters came up to see the exhibit, so I suddenly developed urgent business elsewhere. I had no intention of arguing birth control with three nuns—I know when I'm licked. I'll let you tackle those folks." Allan seems to have done little if any formal teaching, but as Herndon assumed leadership of the medical genetics department he incorporated as much genetics as he could into the medical curriculum. In the first trimester of the 1942–43 school year, he reported, he gave talks to the Bowman Gray Medical Society and the neurology clinic, and lecture courses in pathology, neurology, embryology, and medicine. In 1943–44 he offered a formal course, meeting once a week for twenty-four weeks, to the third- and fourth-year students. In addition, he introduced the third-year students to the "family-history method." He had them take pedigrees on all patients assigned to them on the medical wards. He found the rates of adoption of the method and the quality of the results variable, but on the whole he was pleased to be shaping the course of instruction, ensuring that physicians emerging from the Bowman Gray school would have at least a significant exposure to medical genetics.[41]

Medical genetics at Wake Forest in the 1940s, then, was cast in the mold of Allan's genetic public health. It offered intellectual challenges different from conventional public health, but the essential methods and view of disease were the same: it was a germ-theory-of-genes approach. As Herndon took the reins, the operation moved into the clinic, but it retained its original epidemiological flavor. The focus was always first on the disease, then on the heredity. They never asked, "How does this gene work?" but always, "How is this disease inherited?" What made their work stand out from other efforts in preventive medicine or public health was that, although they always began with the individual patient, their clinical gaze invariably extended to the family, future descendants, and the population. They drew no sharp line between advising an educated woman about the chances of her having a baby with peroneal atrophy or cataracts and sterilizing the defectives of Forsyth County. Reflecting this public health orientation, in 1947 the Department of Medical Genetics moved from Internal Medicine to Preventive Medicine. "The primary aim of medical genetics is the prevention of disease," Herndon wrote in his annual report to the Carnegie Corporation that year. Herndon agreed with Macklin, who agreed with Davenport: prevention of disease in an individual was good; prevention in a family was better; and elimination of the disease was the ideal.[42]

* * *

If Allan's program at Wake Forest was a species of public health and preventive medicine, Dice's at Michigan was a type of human biology. Through the 1940s Dice built a diverse program of research that spanned clinical medicine, anthropology, psychology, and ecology, all of it geared toward understanding the basis of *all* human traits, not just the pathologies. It was, in this sense, quite close to Raymond Pearl's Constitutional Clinic and much closer to Davenport's broad eugenics program than to Allan's narrower one, although Dice attended to the environment far more than Davenport ever did. All of these research efforts, however, shared a commitment to understanding the fundamentals of human heredity with an eye toward human improvement.

The boundaries between vertebrate zoology and medical genetics at Michigan were fluid. In 1942 Dice was promoted to full professor, and the Laboratory of Mammalian Genetics combined with the Heredity Clinic to form the Laboratory of Vertebrate Biology. Although the Heredity Clinic became the largest and most productive branch of Dice's group, ecological and anthropological projects continued to play a role in LVB activities. Members of the group collaborated in different combinations, depending on the project. Heredity Clinic activities included receiving referrals from other medical departments, operating an outpatient clinic, and whatever research was being done by Heredity Clinic staff and associates (members of other medical departments with courtesy appointments in the HC).

Dice had broad, eclectic research interests. He maintained a large colony of *Peromyscus* mice and used them in a variety of studies, both genetic and ecological. He developed a strain with hereditary epilepsy; in 1945 he sent a supply of them to the pharmaceutical company Glaxo for use in testing anti-epilepsy drugs, and the study continued to be funded, largely by the National Institute of Mental Health, well into the 1950s. He also carried out a study of the effects of radiation on mice, in collaboration with Earl Green, at the Roscoe B. Jackson Memorial Laboratory in Bar Harbor, Maine, and funded by the Atomic Energy Commission. His zoological training made it easy and obvious to carry out mouse studies with the aim of human application. Not until the 1960s did this approach catch on widely among human-geneticists.

4.3 The building that became the Michigan Heredity Clinic. Courtesy Bentley Library, University of Michigan

At the same time, he kept his ecological roots, pursuing, for example, studies of geographic variation in wild *Peromyscus*.[43]

Even before the Heredity Clinic officially opened, Dice and Cotterman began collaborating with medical faculty on clinical case studies. In addition to the connections with the orthodontists George Moore and Byron Hughes and the ophthalmologist Harold Falls, Dice and Cotterman cultivated contacts in neurology, medicine, and other departments in the School of Medicine as well as mental hospitals and other institutions outside the university. In 1944 they listed referrals from twelve different sources. When unusual cases came in to these clinics, Dice or, more often, Cotterman would be contacted and they would collaborate to describe the condition and the pattern of heredity. Falls was one of their favorite collaborators. Cotterman and he published a paper on the genetics of ectopia lentis, a displacement of the lens related to Marfan syndrome. Another paper described a surprising hereditary pattern in the Marcus Gunn phenomenon, an abnormal contraction of the pupil used as a sign of optic nerve damage. Moore and Hughes alerted Cotterman when they found any patients with what looked like hereditary dental anomalies. Cotterman published articles, usually with other clinicians, on a wide array of such traits—among them developmental anomalies, heredity of gout, and "status Bonnevie-Ullrich," which a decade hence would be understood to be a sex-chromosome deficiency known as Turner syndrome. Such were the types of studies that led Herndon to deride the early Heredity Clinic research as "anomalies of the eyebrows and minor defects of the teeth."[44]

The Heredity Clinic offered no treatment other than counseling. Early publications reflected a harshly eugenic outlook that took the reproduction of individuals with genetic "defects" as the equivalent of tubercular scoundrels spitting in the water supply. In their ectopia lentis paper, for example, Cotterman and Falls stated, "A kindred of this character frequently becomes a burden to county and welfare funds. This family is no exception." The last line of the paper read, "Eugenic advice is of importance to both the unaffected and the affected members of a kindred exhibiting this dominant pathologic inheritance." In later years, Falls's eugenic judgments softened; by the mid-fifties, he often included sensitivity to the patient's needs and wants as an important determiner of the kind of advice the genetic counselor should give.[45]

Patients admitted to the Heredity Clinic underwent an exhaustive battery of tests and interviews. An eighty-page manual from 1946 documents them, as well as codifying procedure for every clinical and administrative aspect of the Heredity Clinic, down to boilerplate for form letters and specifications for the color and weight of paper to use for different purposes. Typically, someone visiting the Heredity Clinic presented either a medical complaint known or presumed to be hereditary or some other trait of potential genetic interest. This individual was called the "propositus," the term for an individual through whom a family comes medically to light. Considerable care seems to have been taken to secure informed consent as well as permission to interview and examine family members—consent and permission are listed on multiple forms and instructions. The staff used a variety of forms, but among them were "family record" forms, a type of form used also at Wake Forest and a direct descendant of those used at the Eugenics Record

Office at Cold Spring Harbor. The origin of such records, of course, was Francis Galton's anthropological survey back in 1889.[46]

A typical patient work-up by the Heredity Clinic staff involved an updated, medicalized Galtonian biometry. Their parsing of the body relied heavily on a semipopular book published in 1942 called *Family Treasures*. Billed as being directed primarily at amateurs, it was a sort of field guide to the human body, a taxonomy grouped by body part, heavily illustrated with more or less clinical photographs and accompanied by a narrative and descriptive, rather than analytical, text. The staff recorded a battery of test results, including assessments of mental acuity and musical talent, as well as a wide range of anthropometric measurements. They recognized ten different shades of skin color, including four shades of brown but no black. They recorded reflexes, tongue protrusion, "mental condition," "habits," speech, memory, attentiveness, and visual, auditory, and olfactory abilities. In short, they were interested in the same range of traits as Charles Davenport had been, but their measures were less laden with value judgments and stereotypes.[47]

Less laden, but not unladen. "Yellow brown" skin was described as "the usual oriental type," thus simultaneously appealing to folk knowledge about race and lumping people of Asian descent with those of African descent. Subjectivity crept everywhere into the measurements. A submedium depression of the nasion (the point where the frontal and two nasal bones converge) should be recorded, the manual specified, when "a little more than a trace but not the customary depression" was observed. A "+++" forehead slope was defined as having "an exaggerated posterior slope, giving a 'rat face' appearance." Obesity ranged from none, to slightly plump, to average for a person over thirty "or a comparable amount in a younger person," to a "fat individual." Subjectivity in itself is not necessarily insidious, of course, and naturalists have ever sought to partition graded characters into sharply defined categories. It is a necessary procedure whenever one tries to classify objects—to put people, in this case, into bins according to type. Nevertheless, in the murky world where human biology met medicine, the method had inherent moral hazards.[48]

These examination procedures show vividly a key difference between the Michigan Heredity Clinic and the Bowman Gray Department of Medical

Genetics. In contrast to Allan and Herndon's emphasis on pathology, Dice was equally interested in positive human characters and defects. He developed a colorimeter to measure eye color and the tint of the skin, an experiment reminiscent of Francis Galton's eclectic studies of talent and character. He bought a piano in order to test sense of pitch. In the tradition of human genetics and eugenics, by 1941 more than thirty years long, Dice's catholic approach, involving medicine, biometry, psychophysics, and other disciplines, was more mainstream than Allan's and Herndon's more narrowly clinical approach. The zoologist in him was always interested in measuring normalcy at least as much as dissecting pathology.[49] After the war, Dice began to win larger grants to undertake longer-term studies. In 1949 he drafted a proposal for a study of hereditary abilities. Again recalling Galton, Dice's study involved gathering pairs of twins and giving each a battery of psychological and behavioral tests, taking anthropomorphic measurements, and also taking blood for biochemical tests. The experimental question was exploratory rather than hypothetical: the goal was simply an "attempt to discover which . . . abilities are hereditary" and to search for correlations between mental abilities and biochemical or physical traits."[50]

But if the experimental design of the hereditary abilities study was simplistic, it was founded on a strikingly Garrodian axiom. "A fundamental tenet of the science of genetics," Dice began in his proposal, "is that most if not all of the spectrum of genetically controlled variations finds its ultimate explanation in terms of biochemical alterations." It seems to be an extension of George Beadle and Edward Tatum's "one gene, one enzyme" hypothesis, articulated just a few years before, to all of human heredity. Dice obtained a $90,000, three-year grant from the McGregor Fund, a Detroit philanthropy organized in 1925 "to relieve the misfortunes and promote the well-being of mankind." In a 1951 note to C. M. Goethe, one of the enduring patrons of eugenics research and a great admirer of Dice, he described the Hereditary Abilities Study as "part of our general program to measure as precisely as possible the trend of heredity in human populations." At the Michigan Heredity Clinic, Galtonian measurement and population thinking coexisted peacefully with Garrodian biochemical individuality.[51]

It was axiomatic to Dice, as a mammalogist and ecologist, that population principles should apply equally to mice and men. For example, one of the main projects of his group in the 1950s was the Assortative Mating Study. Assortative mating is a basic concept of population genetics that describes the degree to which animals prefer to mate with animals either similar (positive) or dissimilar (negative) to themselves. In nature, positive assortative mating can lead to sympatric speciation: one species splitting into two within a defined geographic area. Thus the implied question in studying assortative mating in humans is: To what extent are marriage choices among urban people leading to the creation of genetically distinct populations? Race is certainly one variable in urban populations, but age, socioeconomic status, education, profession, personal taste, and other variables also enter in.

The project was funded by an "anonymous donor" who gave $100,000 over five years—a huge amount for the day. It was Wickliffe Draper. The principal investigators for the study were Dice and James Spuhler, an anthropologist on the staff. Taking the city of Ann Arbor as a type of natural community, they asked, What is the size of the mating isolate—the group of individuals who interbreed? What is the frequency with which members of the community marry into families with economic and social backgrounds similar to their own? What are the similarities between mates in physical features, mental characteristics, states of health? And finally, what are the effects of these choices on the future structure of the community?[52]

The method was straight out of Galton's *Natural Inheritance*. They would randomly sample the population of Ann Arbor. Each person selected would undergo a wide variety of biometric and psychological measurements—physical traits, such as height, weight, hair color, and so forth, as well as mental traits including IQ, musical ability, and other aptitudes. And each would be extensively interviewed about occupation, income, habits, and of course mating choices. The researchers would then perform statistical analyses to discover correlations among the traits in order to determine whether people who possessed a given trait were more likely to marry someone who shared it. In 1951 the United Press wire service picked up the story of a huge amount of money for research on human mating preferences. The story's lead ran: "The University of Michigan announced the donation of $100,000 to explore the 'decisive question' whether gentlemen actually prefer blondes." The project, the reporter stated, would attempt to discover "why certain types of men and women are attracted to each other." A woman in Germany, skeptical of such a deterministic approach to human mate choice, clipped the story and sent it to Dice, penning sardonically beneath it, "Because they love each other!"[53]

* * *

Cover one eye when looking at the heredity clinics and you think you are looking at an old-fashioned eugenics office. Cover the other, and you see something very like a medical genetics clinic of the "modern" period. Stereoscopically, the images blend, gaining texture and depth. Their founders saw them as the next logical step in the medicalization of negative eugenics. Both Lee Dice and William Allan had strong ties to the Progressive-era eugenicists, and both men explicitly modeled their institutions on the Cold Spring Harbor Eugenics Record Office. Davenport and Laughlin may have been ignominious by the 1940s, but among those interested in human health, eugenics remained a noble, intelligent, compassionate goal. At the same time, the heredity clinics' directors and staff were central in the professionalization of human genetics, beginning with the formation of a professional society in the late forties. They are the direct ancestors of the divisions and departments of medical genetics of the nineteen fifties and sixties, founded by sundry "pioneers" and "father" figures, celebrated in internal histories of the field. They helped ensure that when human genetics became a profession, its practitioners would work in medical schools.

The heredity clinics crystallized the idea that negative eugenics was a medical procedure. Frederick Osborn did not invent this distinction—since the late 1920s, negative eugenics had been increasingly a medical practice—but he was instrumental in institutionalizing it. In the trusted environment of a hospital, concern with immigration and marriage laws gave way to eugenic advice, often dispensed by a physician, and medical sterilization. Once, the unfit were outcasts; now they were patients. No one would have denied the social associations of these unfortunates—that they were the inbred poor of Appalachia, or the descendants of Africans with sickle cell anemia. But the justification for eliminating a tainted human "strain" was now medical, not sociological. Allan's family history method brought negative eugenics from a population level of analysis to the level of the individual. One no longer had to rely on crude techniques such as marriage or immigration restriction to eliminate hereditary disease; one could operate at the level of keeping *this* person from having a baby with a given disease.

The heredity clinic workers saw no conflict between Garrodian concern for the individual and Galtonian concern for the population. Prevention of suffering is prevention of suffering, whether now or in the future. But the urgency of racial improvement was waning. The commitment remained but the zeal had ebbed, pushed back as much by the physician's concern for the patient as by technical or theoretical advances in genetics. Gone were the statistically naïve predictions of huge changes in gene frequencies in a few generations; these researchers had a sense of how long it would take to achieve meaningful change. Many medical geneticists and eugenicists of the late 1930s were becoming more philosophical. So long as the genetic material remained enigmatic and so long as direct intervention in human heredity remained ethically off-limits, they would have to be patient.[54]

Patient, however, does not mean inert. Although the heredity clinic workers understood the limitations of their knowledge, they pressed forward avidly in building and growing their profession. At the end of the war, the optimism about human genetics in the early 1940s would blossom, when the world gained a stunning new sense of the urgency of stemming human genetic deterioration.

5

How the Geneticists Learned to Start Worrying and Love Mutation

ON AUGUST 23, 1939, GERMANY AND THE Soviet Union signed a nonaggression pact—and nearly two hundred Americans and four hundred other scientists convened for the Seventh International Congress of Genetics in Edinburgh. It was to have been held two years earlier, in Moscow, but it had been delayed by politics dimly understood at the time. It was rescheduled for the summer of 1938, delayed again, and planned for 1939. Soviet officials refused to sanction the meeting. Mendelian genetics was becoming highly politicized in a climate of strong Communist Party support for Trofim Lysenko, the agronomist who claimed to have created new species of wheat in a single season through the special treatment of "vernalization." Mendelian genetics said it couldn't be so, and so Mendelian genetics became a political threat. Hermann Joseph Muller, the brilliant but prickly *Drosophila* geneticist, had been working in the Soviet Union and had been much enamored of the Soviet system—he had been on the original organizing committee for 1937—but as a staunch Mendelian he had run afoul of Stalin and fled to Edinburgh in 1937. Now F. A. E. Crew, director of the Institute of Animal Genetics at the University of Edinburgh, offered to step in and host the meeting. Muller would help him organize. By the time the Congress at last came together, the world was on the brink of war. Anxiety suffused the congress from the outset.[1]

One of the most memorable events of the meeting had a Marxist flavor. Before the meeting, Science Service, the Washington, DC–based science education and lobbying group, posed to the delegation the fundamental problem of eugenics: "How could the world's population be improved most effectively genetically?" In response, Muller drafted a "geneticists' manifesto," a scientific counterpart to the revolutionary socialist documents he found so romantic. Human genetic improvement, it said, would be accomplished through inbreeding and selection, carefully managed and carried out in a context of a much more egalitarian and economically secure social system, better reproductive technology for birth control and sterilization, an educated population willing to improve itself and in agreement on the direction that improvement should take, and much deeper knowledge of the mechanisms of heredity. Muller's fellow leftist biologists, including J. B. S. Haldane, Lancelot Hogben, Julian Huxley, Herbert Spencer Jennings, and the biochemist Joseph Needham, were glad to sign on to the cause. Also on the list of signatories were Bronson Price, a population geneticist with the Census Bureau and a friend of Muller's, as well as a fellow Soviet sympathizer; Arthur Steinberg, a former student of L. C. Dunn's; the distinguished population geneticist Harold Plough; Caryl Haskins, the entomologist, human-geneticist, and later president of the Carnegie Institution of Washington; and the physician George P. Child.[2]

As the weeklong congress wore on, focusing on science became increasingly difficult. Steinberg later wrote that buildings were sandbagged and gas masks were handed out to the delegates. The French and the Germans left the meeting early. Sections of the meeting were reorganized to accommodate a truncated timetable and abbreviated roster. Finally, it became obvious that everyone had to leave. Many British and American delegates booked reservations on the British passenger liner *Athenia*. The ship left Glasgow on August 31, stopped in Liverpool, and sailed for New York on September 1, the day the Germans invaded Poland. On September 2 it was torpedoed 260 miles off the Hebrides, the first ship sunk in the Second World War. Several more Americans had found passage on the freighter *City of Flint*, hastily outfitted for passengers. When the *Athenia* went down, the *City of Flint* was near; it changed course and returned to fish survivors out of the cold, gray Atlantic. Among the waterlogged and shivering scientists hauled aboard were Bronson Price and Charles Cotterman. On ship, they saw among their hosts Steinberg, Plough, and the young researcher who would, after the war, emerge as one of the saviors of human genetics: James V. Neel. Neel must have been especially glad to see Cotterman, since they had bunked together in Edinburgh. Later, Muller, still in Edinburgh, received a postcard from his good friend Price: "That torpedo knocked me out of my deck chair but left my fillings intact."[3]

Thus are connections made, networks formed, agendas built: through the happenstances of politics, friendship, and tragedy. Associations formed before the war became key to developments postwar. Muller, already agitating for an enlightened, socialist eugenics, emerged in the late forties as America's new leading monger of hereditary anxiety. And if Neel did not personally rescue human genetics from the sharks, he was certainly a paying passenger on the boat. His career choices and diplomatic style became a model for human genetics in the postwar era. Muller and Neel often disagreed on matters of policy, procedure, and theory. But they shared a commitment to exploiting the opportunities of the atomic age to further the study of human heredity.

The aftermath of the war created a new professional and public environment for human genetics. Suddenly, the relationship of heredity to human health seemed an urgent matter. Pulled along on the current of postwar public interest in science, professional geneticists who had shied away from problems of human heredity in the 1930s were drawn back to them again. Those human-geneticists long interested in human problems seized this momentum to professionalize. Because human genetics continued to be allied with eugenic goals of human improvement and directed evolution, professionalization in human genetics was partly a question of whether and how to institutionalize the social control of heredity. And yet the field's leaders also felt it essential to dissociate human genetics from earlier and now disreputable views. They had to create, in other words, the feel of a fresh, new, and ethically squeaky-clean field—without throwing out the eugenic baby with the Nazi bathwater. By the time the Cold War had fully crystallized, human-geneticists had a professional society, a journal, a new image, and a vital public role. For they were now publicly acknowledged as stewards of the human gene pool—just as Charles Davenport had longed to be.

* * *

Hermann Muller always seemed to be fighting someone. Born in 1890 in New York City, he had participated in the exciting *Drosophila* work in T. H. Morgan's laboratory, informally at first, then, after 1912, as a regular member of the group. Convinced that his fellow students Calvin Bridges and Alfred Sturtevant were taking credit for his work, he had broken with the Boss to strike out on his own. In the late teens, he had already identified the great themes of his career: mutation, human improvement, and socialism.[4]

As early as 1923 he had argued that "the basic mechanism of evolution" is mutation. "And," he continued, "since eugenics is a special branch of evolutionary science it must be equally concerned with this problem." At the time, Muller called this process "mutational deterioration"—a technical specification of the old eugenicist's concern over "degeneration." He also became convinced that genes were material objects—which was far from

5.1 Theophilus Painter, Clarence Oliver, Wilson Stone, Hermann Joseph Muller, and many, many fruit flies, University of Texas, late 1920s. Courtesy of Lilly Library, Indiana University

obvious at the time—and speculated that someday we might be able to grind up genes in a mortar and analyze them. At the University of Texas, he constructed baroque strains of *Drosophila* that allowed him to show at last that X-rays induce mutations in the genetic material, demonstrating at once the material nature of the gene and the social hazards of radiation. And he began developing a "Bolshevik eugenics." His increasing involvement with socialist and communist politics forced him to leave Texas, and in 1932 he found himself in Berlin, working with his Russian expatriate friend Nikolai Timofeéff-Ressovsky. He urged Timofeéff to collaborate with the physicist Karl Zimmer, also in Berlin, to use the combined power of physics and biology to unravel the nature of the gene and of mutation. In 1935, with the young quantum-mechanist Max Delbrück, Timofeéff and Zimmer produced a multidisciplinary paper on the nature of gene mutation and gene structure that is considered a founding paper of molecular biology. Delbrück emigrated to the United States, pioneered the genetics of bacteriophage, and midwifed the nascent field of molecular biology, mentoring many of its pioneers, including the young James D. Watson. Thus, though Muller himself remained steadfastly unmolecular, he catalyzed the development of a molecular approach that, decades later, began to realize his vision of purposive control over human evolution.[5]

Leaving fascist Germany for communist Russia in 1934, Muller finished *Out of the Night,* a popular book he had been working on since the twenties. It made the case for "eutelegenesis," a form of voluntary eugenics combined with what would come to be called in vitro fertilization. The Soviet Union, he imagined, was the ideal—perhaps the only—place to equalize opportunity and implement a meaningful and just program of eugenic selection. However, when Muller presented the book to Stalin, the Soviet leader rejected it, and, Muller believed, ordered an attack against it. By that time, Lysenko was on the rise, and Muller fled. He stopped briefly in Paris and Madrid before securing a position with Crew at Edinburgh. Returning to the United States in 1940, he obtained a position at Amherst College, but soon got in trouble for his political activities once again. Friends worried that the brilliant but irascible geneticist might be unemployable. Just as the war ended, he received what must have been a very welcome offer from Indiana University for a stable, tenure-track faculty position. He leapt at it.[6]

The move to Bloomington put Muller in the epicenter of the new American human genetics. He was within driving distance of Columbus, Ohio, where Laurence Snyder taught medical genetics at Ohio State University. Snyder was the hub of a small but growing network of physicians, geneticists, and zoologists interested in human heredity. Discovering a taste for administration, in 1947 he moved to the University of Oklahoma to become dean of medicine; his close colleague Madge Macklin moved to Ohio State to fill his position. In Chicago, H. H. Newman turned over the reins of their psychological twin research to Herluf H. Strandskov, a student of the superb Chicago geneticist Sewall Wright. In Ann Arbor, Dice was building momentum again with Snyder's student Charles Cotterman and the talented young physician-scientist Jim Neel. To the west, Eldon J. Gardner opened a new institute at the University of Utah. Muller's former student Clarence P. Oliver had left the Dight Institute in Minneapolis to begin a small human genetics program at the University of Texas in Austin. After a short hiatus, the zoologist Sheldon C. Reed took the reins at the Dight—and promptly concluded arrangements with Cold Spring Harbor to transfer the bulk of the Eugenics Record Office data to Minneapolis, thus completing the shift of eugenics and human genetics out of New York to the upper Midwest. Snyder, Cotterman, Dice, and Strandskov, in particular, drew Muller into their circle as they sought to professionalize American human genetics.

Although much had changed institutionally in the 1940s, technically the work was much the same as it ever had been. At the dawn of the atomic age, American human genetics research would have been easily comprehensible to Charles Davenport, Henry Goddard, or David Starr Jordan decades before. Researchers were comparing twins, collecting pedigrees, seeking Mendelian patterns, and trying to calculate the proportion of heterozygotes in a population. The prewar optimism over human genetics stemmed mainly from two things: the field's recent institutional growth and the explosion of knowledge of the genetics of blood groups.

The blood group field had grown so much, in fact, both in basic knowledge and clinical application, that it was an utter mess. Back in 1941 Strandskov had written, "In no field of science is there greater lack of uniformity with respect to the usage of symbols and greater lack of adherence to conventional genetic rules than in the field of human genetics." The crux of the problem was the blood groups, and the crux of blood groups was Rh, the rhesus complex.[7]

The accepted scheme for classifying Rh alleles was an abominable mash-up of two different systems. One, using cozy and familiar subscripts of letters and numbers, had been proposed by the New York physician Alexander Wiener. The other, austere and scientific with superscript primes, had been developed by the English geneticists Robert Russell Race and G. L. Taylor. Though the system was messy, physicians were comfortable with it and saw no point in changing; Wiener, further, was stubbornly convinced of the virtues of his system and loath to let it go. The geneticists, led by Strandskov, were appalled by the sloppiness of the system and sought to reform it to make it consistent and logical. By the mid-forties, the Rh alleles had become a biomedical Gaza Strip—a tiny bit of terrain claimed by two communities forced to live together by events larger than they.[8]

The geneticists pressed for summit talks. In 1945 Snyder invited his colleagues to form a national committee on blood group nomenclature to sort out the mess and thereby enable the field to get some traction once again. Among them was Herluf Strandskov. Since before the war, he had been agitating for various measures to professionalize human genetics. As long as they were systematizing human gene nomenclature, he suggested, they ought to do the entire set of human genes and be done with it. The committee might also consider pedigree symbols and questions of the "integration of research in human genetics." Further, other countries were adopting their own systems. British human-geneticists had formed a similar committee. They insisted that human genetic nomenclature was a unique problem that "should be recognised as totally different from that of any other animal or plant." A Canadian committee, however, wondered what all the fuss was about. "We feel," they wrote, "that the nomenclature of human genes is, with few exceptions, developing in reasonable accord with the usage most widely accepted in genetics, and that no real crisis in nomenclature has yet arisen." The nomenclature committee thus put a dog in an international fight; the dry topic of naming genes was a wedge, a procedural tool for promoting the professionalization of human genetics.[9]

Snyder convinced an apparently reluctant Muller to join the nomenclature committee. He and Strandskov "dragged in Muller by the heels," Snyder told the historian Dan Kevles in 1983. Doing so was a coup for the committee. "We got him in there to give some credibility to the whole thing," Snyder said. Getting Muller on the committee also helped integrate him into the little network of North American human geneticists. Muller shared the other men's belief in the importance of basic research in human genetics and agreed that negative eugenics was the necessary first step but that positive eugenics was the ultimate and more important goal. A subtle but significant change had occurred in eugenic theory in the preceding years. As negative eugenics became medicalized, it came to take priority over positive eugenics—not because it was seen as more important but because it was more practical. Allan had written in 1936, "The only way to attain the goal of positive eugenics is to actually practice negative eugenics." This idea lay at the heart of professional human genetics, and during this crucial decade or so, Muller was its most expressive and trusted exponent.[10]

Snyder received Muller's acceptance of nomination to the committee on August 3. On August 6, the Enola Gay dropped its payload on Hiroshima. Before the bomb, Muller's main concerns had been the accumulation of spontaneous mutations and background radiation, and sources of man-made mutation such as X-ray machines. But when the bombs were dropped, he wasted no time integrating this powerful new source of mutation into his message.

In November 1945 Muller gave the Pilgrim Trust lecture, part of a distinguished but short-lived series hosted alternately by the Royal Society of London and the U.S. National Academy of Sciences. He began by rehearsing his long-standing argument about the dangerous accumulations of mutations. "Probably," he wrote in the published version, "the great majority of persons possess at least one recessive gene, or group of genes, which, had it been inherited from both parents, would have caused the death of the given person." This Muller dramatically called "genetic death." If selection against such recessive lethal alleles were relaxed and people were allowed to "live and breed without limit," he wrote, "their number would creep up and it would be a case of treating everyone for everything." And this was only the baseline level of mutation. The situation would be made worse by "injudicious X-ray treatment, exposure to artificial radioactivity or to special chemicals, or unwise average age of parenthood." Muller speculated that such factors could increase the mutation rate by a factor of three or more, with a corresponding increase in the number of genetic deaths. "Artificial radioactivity" from atomic bombs or other sources made a eugenics program urgent. Like Francis Galton before him, Muller hoped that eugenics would be voluntary. Further, he cautioned against being rigidly hereditarian about genetic improvement. "The costly lesson taught us by the terrible Nazi perversion of genetics," he wrote, ought to serve as a reminder to pay attention to the effects of the environment and to think carefully about what qualities were important to foster in the human species.[11]

The atomic age inspired dramatic religious imagery and references. On witnessing the Trinity test in the New Mexico desert, Robert Oppenheimer was awed by the power of physics: "Now I am become Death," he said, invoking the Hindu god Siva, "the destroyer of worlds." Muller, reflecting on the power of atomic-age genetics, preferred biblical images of purity and perfection. "Mankind," he wrote, "is cursed or blessed with what has been called 'the divine discontent,' which drives him ever further everywhere."[12] His positive eugenics was an expression of that divine discontent—framed by scientific knowledge of the gene. It was an impulse to mold the intellect and engineer the soul in an ever-ascending curve toward perfection.

After the war, Muller became America's most famous biologist. In October 1946 the Karolinska Institute in Stockholm announced that he would receive that year's Nobel Prize in Physiology or Medicine "for the discovery of the production of mutations by means of X-ray irradiation." In his banquet speech, Muller linked his old argument about genetic load to the new threat of atomic radiation. "So long as we cannot direct mutations, then, selection is indispensable," he wrote, and, since most mutations are harmful, "progress in the hereditary constitution" can be made "only with the aid of a most thoroughgoing selection of the mutations that occur." He repeated his call for protection of the gonads during known exposure to radiation, such as medical X-ray exams. And further, "with the coming increasing use of atomic energy, even for peace-time purposes, the problem will become very important of insuring that the human germ plasm . . . is effectively protected from this additional and potent source of permanent contamination." The bomb, in other words, had ratcheted up the need for eugenics manyfold.[13]

* * *

Twenty-five years Muller's junior, Jim Neel became one of his principal nemeses. Born in 1915 in Hamilton, Ohio, James Van Gundia Neel grew interested in genetics while an undergraduate at Wooster College in the thirties, where he worked on *Drosophila* under W. P. Spencer. He went on to a Ph.D. in genetics at the University of Rochester, where he became the first American student of the German expatriate Curt Stern. During the war, Stern, it turned out, had been doing experiments on exposure to radiation as part of the Manhattan Project. Rochester had been a center for biomedical research related to atomic radiation. Fresh Ph.D. in hand, Neel had treated himself to an Atlantic passage and attendance at the seventh international genetics congress, which ended with that dramatic rescue aboard the *City of Flint*. He spent the academic year 1941–42 at Columbia University on a National Research Council fellowship, where he worked with L. C. Dunn and the Russian transplant Theodosius Dobzhansky; the latter would also do battle with Muller. On December 4, 1941, just before Pearl Harbor, Neel applied to the medical school at Rochester. In his memoir, he wrote that he figured that in a war, a physician would be more useful than a geneticist. But in 1942 he visited the remains of the Eugenics Record Office. Simultaneously impressed by the potential power of medical genetics and the uselessness of much of the ERO data, Neel determined to help rebuild the field. "It was a real gamble to believe I could bring the rigor of *Drosophila* genetics into this arena," he wrote. Medical training was part of that rigor, although medicine was a larger part of Neel's identity than his training. Under the military's accelerated program, he got a medical degree in two years, finishing in September 1944 and then staying at Rochester for an attenuated internship and residency. Neel was never very interested in clinical work; his M.D. was a credential to do human biology.[14]

Lee Dice had been looking for someone like Neel; for years he had understood that an ideal medical geneticist would have a Ph.D. in genetics and an M.D. Searching for Herndon's replacement at the Michigan Heredity Clinic, he had considered Ray Anderson, who had gotten a Ph.D. doing *Drosophila*

5.2 James V. Neel, who converted the Michigan Heredity Clinic into the Department of Human Genetics. Courtesy of Bentley Library, University of Michigan

genetics under Pete Oliver and was now working toward an M.D. with a concentration in pediatrics. Anderson, however, chose to stay at the Dight. Cotterman suggested his 1939 bunkmate and savior Jim Neel.[15]

Though young, Neel was supremely confident and authoritative—there was little doubt that he could administer the medical activities of the Heredity Clinic. However, he was becoming increasingly involved with the largest human genetics experiment ever done. As soon as the Hiroshima and Nagasaki bombs were detonated, radiobiologists began to organize to study the effects. By the end of the year, a preliminary survey of the Japanese survivors was completed. In 1946 a research team assembled by the National Academy of Sciences–National Research Council began feasibility studies for a longer-term project; Neel was involved. In the summer of 1946 President Truman signed the Atomic Energy Act, which transferred control of nuclear materials from military to civilian hands and established an Atomic Energy Commission to manage them. The AEC would be the principal source of funding for human genetics research in the atomic age. Among its projects would be an Atomic Bomb Casualty Commission. In the fall Neel and others made a reconnaissance trip to Japan; in January 1947 Neel went to Washington, DC, to present a plan for studying the genetic effects on blast survivors.[16]

His research strategy was to seek evidence to refute a "null hypothesis" that no long-term genetic damage had occurred to the survivors of the blasts. No Japanese would be treated for any observed or anticipated effects. Thus science trumped medicine; the plan relinquished the humanitarian high ground in exchange for rational experimental design. Neel argued for his plan as logically sound and scientifically conservative, as taking advantage of a unique research opportunity. Further, he thought it politically important for science to help quell public anxieties.[17]

Muller thought anxiety was the rational response to this new genetic threat. He worried that the study was designed to mask real damage and hence real risk of future harm. His years of work generating mutations with radiation had persuaded him that a) genetic damage from the bombs was a virtual certainty and b) it could well be difficult to detect. Sensationalist media accounts whipped up fear about atomic genetic monsters, but to Muller, the real danger was genetic death by a thousand cuts, the accumulation of many small, perhaps individually undetectable mutations that collectively would lower fertility, vigor, and overall human quality. It was the old eugenicist's argument about degeneration, amplified by atomic radiation. Muller argued for presenting a skeptical view of the genetics study, so that if it found no evidence of genetic damage to the Japanese, the public would not consider that the last word on the subject. In the end, the committee approved the project with these unenthusiastic words:

> Although there is every reason to infer that genetic effects can be produced and have been produced in man by atomic radiation, nevertheless the conference wants to make it clear that it cannot guarantee significant results from this or any other study on the genetic material. In contrast to laboratory data, this material is too much influenced by extraneous variables and too little adapted to disclosing genetic effects. In spite of these facts, the conference feels that this unique possibility for demonstrating the genetic effects caused by atomic radiation should not be lost.[18]

It was the Cold War version of Lewellys Barker's 1927 "polyhybrid heterozygous bastards" argument. Humans are too complex, their social structures too elaborate, the obstacles to gathering complete sets of reliable data too great for field data to be trustworthy. From a scientific point of view, the Japanese bomb survivors presented an extraordinary "natural experiment," an opportunity that would be irresponsible to squander. But natural experiments are notoriously weak in their controls.

Neel had the ear of the government, but Muller had that of the public. He took advantage of many speaking opportunities to spread his message of the genetic dangers of radiation. In April, for example, he gave a version of his Nobel speech in a plenary lecture for a conference on public health at the New York Academy of Medicine; it was picked up by the *New York Times* with the headline "Radioactive rays held peril to race." The next month, he gave a talk at the University of Illinois Medical School on "Human Erosion by Mutations." Herluf Strandskov came from the University of Chicago to hear it. Afterward, he recruited Muller to his cause. "I enjoyed your lecture," Strandskov wrote in a follow-up note. "I am convinced, as I know you are, that something must be done sooner or later to prevent the piling up of detrimental genes in human germ plasm." Getting down to business, he wrote that he would be in Bloomington on Saturday, June 7. Could they have lunch together? "What I should like to talk to you about is the possibility of some informal meeting (at the Christmas meetings) at which we can discuss certain human genetics problems."[19]

The "human genetics problems" Strandskov wanted to discuss were professional rather than scientific. Over lunch, Strandskov secured Muller's cooperation to participate in a rump session on human genetics at the end of the upcoming AAAS meeting, always held over the week between Christmas and New Year's. The main problems were four. Snyder would handle nomenclature. Strandskov himself would discuss the need for a professional society. Charlie Cotterman, whose famous "non-publications" were ironically symptomatic of a meticulous and perceptive editorial eye, would lead discussion on publication of results in human genetics and the need for a journal. Strandskov asked Muller to address problems of "international agreement and cooperation." The phrase was one of the terms that had replaced the prewar language of social control. It was partly euphemism and rhetoric, but also partly a more accurate description of postwar genetic policy. For geneticists—and for Muller in particular—it carried heavy connotations of "fighter against Lysenkoism." Lysenko's dogma had plagued Soviet genetics since the 1930s, but it gained new vigor in the postwar era, as tensions between the Soviet Union and the West mounted. Science and technology became symbols of national pride; much of the Cold War was fought in laboratories. Physics and engineering, of course, had the highest profile, but biology was becoming increasingly important. "Mendelism-Morganism" became a political affiliation as well as a scientific stance. A meeting of the Lenin All-Union Academy of Agricultural Sciences, scheduled for Moscow in the first week of August 1948, was a high-water mark for Lysenkoism. The meeting took place shortly after the Eighth International Congress of Genetics in Stockholm. Muller was the president of the Stockholm conference. Using his status as Nobel laureate, and still carrying a grudge against Stalin for rejecting *Out of the Night,* Muller attacked Lysenkoism as pseudoscience and sought to expose the ouster and even death of distinguished Soviet Mendelians. Under Muller, the new Society would be a bulwark against the corrosive forces of Lysenkoism and, more generally, of genetics misconstrued for political ends.[20]

After the fall AAAS meeting, Strandskov sent a questionnaire to his colleagues about the desirability and structure of a new society for human-geneticists. By mid-December he had some eighty returns, almost all of which were positive. That winter, he, Strandskov, Snyder, and Cotterman met and decided to press ahead and call into existence the "Human Genetics Society of America," so that they could begin naming officers. This they did.

* * *

The human-geneticists seized the moment of perhaps the strongest support for science in American history. It was widely believed that science had won the war—with the atomic bomb, with radar, with computing and cryptography, with penicillin and vaccines. In 1945 Vannevar Bush's report *Science: The Endless Frontier* spelled out a vision of science as vital for keeping the peace and as a potent tool for postwar dividends in quality of life during peacetime. Bush recommended such bold policy steps as the creation of a federal science agency—this became the National Science Foundation in 1950—and big increases in support for science education. Physics, of course, was the most glamorous science of the day, but developments in biology—especially bacteriology and virology—were advancing rapidly. In the late 1940s, geneticists, microbiologists, radiobiologists, and biological physicists capitalized on increased funding for research from federal agencies, especially the Atomic Energy Commission and the Department of Defense.[21]

Biologists took a cue from the physicists and became far more visible politically. A loose coalition of politically liberal biologists increasingly spoke out on questions of scientific cooperation and, especially, race. L. C. Dunn, the longtime adviser to Frederick Osborn, and Theodosius Dobzhansky were among the most outspoken. Ironically, they often took a stance in opposition to Muller; the erstwhile communist was increasingly heard as a voice of stasis and uniformity, and therefore conservatism. In 1950 the United Nations Education, Scientific, and Cultural Organization (UNESCO) put out a controversial statement on race, with a follow-up and clarification by geneticists and anthropologists a year later. The biologists asserted that race was simultaneously a biological fact and a social myth—they wrestled sincerely but somewhat naïvely with the implications of this dichotomy. The conferences produced the term "ethnic group" to replace "race" in social contexts, although not all geneticists considered this helpful. Most striking was the argument—advanced primarily by the anthropologist Ashley Montagu—that biology "proved" the nonexistence of race and even the primacy of cooperation and brotherhood in human nature. The historian Will Provine later interviewed many of these researchers and found that biological dogma on questions such as the effects of race crossing was based not on new data so much as on a changed political climate; the evidence was the same as before the war, but the scientific truth had evolved with the times. These activist-scientists joined committees, held conferences, made reports, and received grants from agencies such as the Atomic Energy Commission and the National Academy of Sciences–National Research Council. Muller's student H. Bentley Glass, in Baltimore, seemed to be on every committee all at once—from the AEC Committee on the Genetic Effects of Biological Radiation, to the BEAR committee, to the Pugwash Conferences on Science and World Affairs. From mutation to race to disease to population growth, human heredity was increasingly seen as fundamental to human problems and life in the atomic age.[22]

In this setting, the formation of a professional society concerned with the stewardship of the human gene pool seemed of vital importance, even of national security. Yet the field's still-dodgy reputation freighted even mundane matters such as membership policy with significance.

The founders faced two main problems with regard to membership. First, the eugenic fringe. A new society needs a big membership to be viable, yet, Muller observed to the rest of the group, "accepting all applicants might imperil new society swarming it with amateurs, cranks, propagandists, like woman who spoke at our meeting." Available documents do not record the identity of the woman, but she was of a type: "crank" and "propagandist" were euphemisms for eugenics zealots who always seemed to come out of the woodwork to call for the extermination of some social or ethnic group. The organizers, and indeed nearly all human-geneticists of the time, self-consciously distanced themselves from the propaganda and ideology of earlier eugenics movements, yet they had no desire either to repudiate eugenics entirely or to reject the substantial funding that wealthy eccentric eugenicists could offer a fledgling profession.[23]

Even after filtering out the amateurs, cranks, and propagandists, the first roster of members in August 1948 included Wickliffe Draper, the cagey white supremacist philanthropist; Samuel J. Holmes, the West Coast anthropologist and author of the 1924 *Bibliography of Eugenics;* Frederick Osborn, heir apparent to the eugenics movement, and racist anthropologists such as Carleton Coon and Earnest Hooton. During the war, Hooton had written that the forbidden fruit was knowledge not of sex but of human heredity and that "we had better get our teeth into this apple, . . . spit out the worm-eaten parts, and chew the rest. Otherwise, we shall presently have no teeth left and the already ravaged Garden of Eden will be overrun by morons, criminals and atavistic brutes." There remained, then, a significant faction of charter members—many of whom wielded considerable financial or intellectual power—for whom human genetics represented their hope for racial purity. The Society held its collective nose and welcomed them in.[24]

The second membership problem was physician-scientist relations. Most of the founders identified primarily as geneticists; to what extent should they extend a hand to the medical community? Neel, the M.D./Ph.D., began to be involved with the administration of the society, and he joined vigorously in the membership discussions. "First of all," he wrote to the others in September 1948, "I feel strongly that we should at the outset strenuously endeavor to increase the membership of medical men in this society. The bulk of the work in medical genetics will be done by medical men." Cotterman apparently agreed, but Muller was opposed to granting M.D.s full membership. Neel's position was made more difficult by the fledgling organization's early alliances. Like many startup societies, they started as a satellite of an established and much larger organization, which would provide a stamp of recognition, a venue for meetings, and an initial pool of members. The human genetics society hitched its wagon not to the AMA or any other medical society but to the American Association for the Advancement of Science (AAAS).[25] Muller would appear to have won that battle.

On September 11, 1948, the HGSA had its first meeting, in Washington, DC, as a satellite to the annual AAAS meeting and supported by the AAAS. The AAAS did have some M.D. members, but they were a minority and tended to be physician-scientists more than clinicians. They convened on the campus of George Washington University in Foggy Bottom and at the Federal Security Auditorium, two miles from the National Air and Space

Museum. The program was short: it was essentially a business meeting with a lecture and social hour, but the society was launched.[26]

The highlight of the meeting was Neel's plenary lecture, on one of the classic problems of human genetics, "The detection of carriers of hereditary disease." Carrier detection was a problem for geneticists trying to distinguish different genetic constitutions with the same phenotype. For physicians, it was analogous to the problem of asymptomatic carriers of infectious disease. And it was a problem for eugenicists worrying about undetected genes spreading through the population. The detection of carriers has always been eugenicists' key problem: the prospect of undetected carriers of putatively single-gene conditions such as feeblemindedness gave the night sweats to Harry Laughlin, Henry Goddard, and others of the Progressive era. Neel's talk thus played well before the wide range of interests in the audience: the hard-nosed mathematical geneticists who liked a good tough problem, the handful of medical men seeking to apply genetics in "real-world" situations to relieve suffering; and the eugenically minded anthropologists and others who could see it as getting to the heart of the problem at last. Near the end, Neel remarked on the breadth of appeal of his subject. He acknowledged the physical anthropologists, the physicians, and the mathematical geneticists. He even recognized the eugenicists, albeit coyly, with reference to the "sociological implications" of detecting a high proportion of genetic carriers of disease. "For with this detection," he concluded, "there arise new lines of approach in the field of preventive medicine, and the sociological consequences may be far reaching." Neel did not spell out those sociological consequences, but they of course would revolve around reproduction. Who should be allowed to have children? How do we estimate genetic risk? Should genetics shape social policy?[27]

The society's second meeting was better fleshed out. Held in New York at the end of 1949, it was, like the first, a satellite to the big annual AAAS meeting. With fifteen months to plan, the organizers were able to assemble a full program of scientific papers. It was a representative cross-section of problems in human genetics. Traditional modes of analysis from the old eugenics days continued to predominate. D. C. Rife, another Snyder student, presented a biometric study of whether you could tell Catholics from Protestants from Jews based on height, weight, or gross head dimensions. The Utah group led by F. E. Stephens and Eldon Gardner compared pedigrees. Clarence Oliver, now in Texas, presented his studies on twins. Alexander Wiener spoke about blood groups and nomenclature. Race (Jim Neel and colleagues), sex (I. J. Greenblatt and L. Gitman), and mental deficiency were touchstones, as ever. Eye diseases (Harold Falls of Michigan) had always been a major category of hereditary disease, and still were. Most of the diseases examined were simple Mendelian ones, but what we now call complex genetic disease was also well represented, with obesity (E. L. Reynolds), diabetes (Arthur Steinberg and R. M. Wilder), breast cancer (Oliver), and such mental diseases as "manic-depressive psychosis" and schizophrenia (Franz Kallmann). On the last evening, the organizers made a gesture to the eugenic origins of the field by holding a "Biologists' Smoker" at the American Museum of Natural History, home of the Galton Society.

* * *

At this second meeting, Muller gave the society's first and grandest presidential lecture; it set the tone for two decades' worth of thinking on heredity, mutation, and society. The published version, "Our Load of Mutations," is a sprawling eugenic tour de force and his most important and widely read essay. Although Muller was always primarily interested in hereditary improvement, like Frederick Osborn, William Allan, and other contemporary eugenicists, he understood that negative eugenics—the prevention of defectives—was the practical first step. Pragmatically, they all recognized that such negative eugenics was both technologically more feasible and socially more acceptable than positive eugenics. Indeed, the atomic bomb allowed Muller and others to create a sense of urgency that eugenicists had missed in the recent several decades. And the president's podium at the ASHG meeting gave Muller a pulpit. From it, he preached a hellfire-and-brimstone sermon that recalls earlier arguments about genetic purity (see chapter 2).

The essay begins by taking aim at physicians, citing what he calls the "prevailing view," quoted from a 1947 essay in *JAMA*, that mutation as a direct cause of disease is extremely rare and, further, is difficult to detect; thus the conclusion that medical men need not pay serious attention to genetics. Muller rebutted that mutation was in fact both detectable and far more common than widely believed. Using statistical arguments, he estimated the number of genes in man, the rate of mutation per gene (borrowing values from *Drosophila*), and the average selective disadvantage of a given mutation. He estimated that the average human being had eight mutations. Because, as Muller had always believed, the great majority of all mutations are disadvantageous, these eight mutations conferred a 20 percent genetic disadvantage over an idealized "all-normal man." That meant, he wrote, that "most of us have a nearly 20% chance of death or of reproductive inefficacy, from genetic causes." This was the "genetic load," a percentage measure of an individual's distance from genetic perfection.[28]

Although he saw the nobility of medicine's humanitarian life-saving impulse, he found poignant the conflict between individual and population that lay at the core of medical genetics. "We must admit," he continued, "that modern methods do result in the saving for reproduction of many mutant genes which otherwise would have been eliminated by the defects they produced." The suggestion that medical technology would compensate for this gradual genetic deterioration was, he wrote with ecclesiastical imagery, "as effective as trying to push back the flowing waters of a river with one's bare hands." There was exactly one way to avoid "the fiasco of a full fledged resumption of ordinary natural selection," namely "purposive control over reproduction," taking control of our own evolution, refusing to leave to nature and chance the genetic choices that are made with each conception.[29]

Muller was adamant that there should be no coercion: controlling our evolution must occur "through the freely exercised volition of the individuals concerned, guided by their recognition of the situation and motivated by their own desire to contribute to human benefit in the ways most effective for them." It was, then, a Galtonian eugenics, nestled in a bed of socialism. But he was realistic about the social evolution that must accompany the project. "A deep-seated change in mores would be necessary," he admitted, as well as "a far more thoroughgoing and widespread education of the public in biological and social essentials." Muller, increasingly drawn into contact with doctors, recognized the level of cooperation required for systematic human improvement. Voluntary reproductive control, he wrote, is the "necessary complement to medicine and all other 'euthenic' practices."[30]

A believer in top-down solutions, Muller took it for granted that effecting this change should be a government educational project. But the opposite seemed to be happening: he believed that government agencies were using propaganda to minimize public motivation for eugenic change. Drawing on his experience with the ABCC, he argued again that the Japanese data would probably show no effect of radiation on the children of blast survivors. He recognized that such "negative" findings had public relations value for the Atomic Energy Commission: "Assertions have in fact been made that if positive results are not found there, this will have a salutary influence in quieting public fears concerning the genetic dangers of radiation." In Muller's view, this was simply nurturing complacency at a moment of rapidly escalating danger. Humans were being exposed to radiation from many man-made sources, some of which were medical: fluoroscopic screening, ovary radiation to rupture Graafian follicles and induce ovulation, and irradiating the testes as a long-term form of male contraception; use of penetrating radiation and radioisotopes in medicine; X-rays in shoe stores; industrial applications; and of course atomic energy, whose use and the use of its by-products "is only at the beginning of a great process of expansion." Educating people about these risks, he wrote, was literally of vital importance, "one of the first duties of those who appreciate the significance of genetics in human affairs."[31]

In the long term, for Muller as for nearly all eugenicists before or since, mental traits were the ultimate target. "Most of us will agree," he asserted, that "it is the world of the mental life that counts by far the most, the rest being pretty much subsidiary." If a little intelligence is good, more must be better. Looking forward to a positive eugenics, he wrote, "Greater intellectual capacity, and along with it kindlier natural feelings, are surely the greatest biological needs of all humanity." Muller was even willing to accept some physical infirmity if it piggy-backed on mental improvement; recombination would uncouple them eventually, he said, and the negative traits could be winnowed out. For these and many other reasons, Muller concluded that although genetic mutations may not be a direct cause of disease, they are of vital importance to our health. To call mutation a negligible cause of disease in man, as *JAMA* had done, was foolishly blinkered. "None of us can cast stones, for we are all fellow mutants together."[32]

Muller's atomic rhetoric of genetic purity touched off a rockslide of similar arguments and warnings. Acolytes such as the population geneticists James F. Crow and Newton Morton at Wisconsin carried on his view of an ideal genetic man diminished by his load of mutations. The Muller-Crow-Morton view became known derogatorily as the "classical" position—a term intended to connote stuffiness and obsolescence. Theodosius Dobzhansky, a Russian who had emigrated to the United States in 1927 to study *Drosophila* population genetics under T. H. Morgan, gave the position its ironic name. Tarring Muller's position as hidebound and reactionary, Dobzhansky argued that genetic diversity, not homogeneity, was the healthiest condition for a species. He painted his view as "balanced," that is, liberal and open-minded. Though the two men feuded through the 1950s over the nature of mutation and selection, they agreed that atomic radiation made human control of evolution necessary. In a 1958 book, *Radiation, Genes, and Man,* written with his student Bruce Wallace, Dobzhansky asserted that radiation safety measures, such as the isolation and shielding of nuclear reactors, were often prohibitively expensive. The "practical man," wrote Dobzhansky and Wallace, was likely to demand "relatively stern eugenic measures rather than expensive safety procedures." One must be careful, they continued, to avoid the "folly" of earlier eugenic policies, yet, faced with the responsibility of caring for people with genetic disorders, society was likely to demand "the authority to prescribe rules of conduct that will tend to keep the incidence of inherited disorders as low as possible. . . . Involved is some type of regulation of human reproduction and family life." These passages were underlined in ruler-straight red pencil in Bentley Glass's copy of the book. Glass, who had done his Ph.D. with Muller and a postdoctoral fellowship with Curt Stern (Jim Neel's Ph.D. adviser), had eugenics views closer to Dobzhansky's than to Muller's. Writing in 1955 in *Johns Hopkins Magazine,* he discussed eugenic methods candidly, considering the advantages of institutionalization, sterilization, and birth control as negative eugenic measures, rejecting the application of cattle-breeding methods, for example, as socially repugnant, and settling on a more Galtonian program such as a tax system inversely graded to family size. Glass ended wistfully, remarking that the geneticist, like the physicist, had more power than wisdom. He urged caution and care: eugenic progress would come in time.[33]

Among the most moderate voices in American eugenics in the fifties and sixties was that of Frederick Osborn, who kept himself backstage as the secretary-treasurer, although he effectively led the American Eugenics Society. Taking a moderate stand allowed Osborn to boost the society's membership and recruit the support of leading scientists once again. Glass, for example, joined the organization in 1958. Dobzhansky contributed the foreword for Osborn's *The Future of Human Heredity* (1968), writing that "eugenics has a sound core"; it merely needs to be pursued judiciously, fairly, and knowledgeably. Osborn could not have agreed more. In contrast to Muller's socialist eugenics, Osborn mounted a case for a democratic eugenics, founded on principles of equality of opportunity and a free market. Under Osborn's leadership, the American Eugenics Society moved ever closer to the position of professional human genetics.[34]

* * *

Cotterman, meanwhile, had been working to assemble a journal. The founders understood that the journal was critical to the society's success. In practical terms, receiving it would be the main reason for many to become members. Further, it would become the voice of the society, articulating better than policy statements or press releases the society's policy and stance on complex issues. Publishing a house organ is a standard way to consolidate a community around a set of standards or principles and to advertise. The debates and negotiations over the journal of the ASHG illustrate the tensions between M.D.s and Ph.D.s as human genetics professionalized, as well as the founders' artful crafting of a clean professional reputation as they did the dodgy back-room deals that are sometimes necessary to keep a new venture in the black.

Whereas Neel cared enough about doctors to earn a medical credential, Muller remained an animal geneticist at heart; he saw physicians as a conquest. When, in 1948, Neel had suggested to Muller that they make a concerted effort to recruit M.D.s to the new society, he proposed that they use the journal to demonstrate their commitment to medicine. Though Muller agreed, his phrasing suggested colonization rather than cooperation. "The importance of showing medical men that we mean business in their field," he replied, "is such that I thought it should be made obvious to them that we do have a medical editor in an important place in the Board of Editors." Muller suggested Neel himself as the medical editor of the new journal; Neel declined, not unreasonably citing overextension.[35]

The founders needed seed money to launch the journal. Membership dues would never cover the cost. The society did have a small pool of money from a few life memberships they had sold. But Dice urged Strandskov not to squander the organization's only buffer on the journal. Don't spend the monies from the patrons and life members, he urged; invest it and spend only the income. Dice wanted Strandskov to apply to a foundation for a grant. Frederick Osborn found the money. He had been instrumental in the reorganization of the Eugenics Research Association, the outreach arm of Cold Spring Harbor's Eugenics Record Office, as the Association for Research in Human Heredity, which specialized in negative eugenics and medical genetics. Osborn arranged for the ARHH to contribute $2,400 to the society toward starting up the new journal. The journal would in fact launch, although they recognized that it would not take long to burn through that first bit of cash; more funding would be needed shortly. But for the moment they could press ahead.[36]

Cotterman was a perfectionist: the same quality that kept him from publishing his own work made him a meticulous, if slow, editor. His fussiness began with the journal's title. Deeply concerned about its role in the diplomatic relations with M.D.s, Cotterman suggested calling it *The Journal of Human and Medical Genetics.* "It is rather long," he admitted. "But," he continued, "I do feel that the word MEDICAL will aid considerably in accomplishing one of the Society's aims, viz. getting genetics to medical people, and also will have some sales value. Putting the words THE JOURNAL OF in front of HUMAN AND MEDICAL GENETICS seems to me to reduce somewhat the sense of redundancy in HUMAN AND MEDICAL." Cotterman wanted the medical men to *feel* included. He was shouted down by colleagues arguing that, logically, human genetics subsumed medical genetics, so it was unnecessary to be explicit. His proposal provoked a searching philosophical debate. Was human genetics inherently medical? Did human genetics subsume medical genetics? Snyder thought so. He preferred the simpler *Journal of Human Genetics.* "I think that everyone, including physicians and librarians, know [*sic*] by now that human genetics includes much that is medical." A. J. Carlson professed to be "both perplexed and confused" by the proposal. "If the term human genetics does not include clearly all the aspects of the genetics of man in which the medical phases of health and disease are involved, then I do not understand either genetics or the English language." Such rationalism masked a latent chauvinism. "Human genetics" was a biologist's term; it differentiated the Ph.D.s from the M.D.s. To be fair, human genetics did include nonmedical areas such as anthropology and psychology. But like the association with AAAS rather than the AMA, this small gesture reflected a victory for those who saw medicine as an application of science, rather than science as the handmaiden of medicine. *American Journal of Human Genetics* it was.[37]

Cotterman designed the cover, and then he raised another symbolic issue: the matter of a frontispiece or dedication. In a January 1949 note to Strandskov, Snyder, and Muller, he floated the idea of using a portrait of William Allan. It can only have been an act of diplomacy; the two men were intellectual opposites. As a geneticist, Allan was as homespun and practical as Cotterman was sophisticated and abstract. Allan often chastised academic geneticists such as Cotterman for being out of touch with what he considered the real world. For his part, Cotterman was bored with the informal, subjective style of medical geneticists such as Allan. Muller was unfamiliar with Allan—a sign of how removed Muller was from the medical side of things. Cotterman had other suggestions: perhaps Charles Davenport or Dr. Barbara Burks—during the 1930s, she was one of the most active scientists at the Eugenics Record Office—although he noted that the *Journal of Heredity* had recently given them similar honors. Francis Galton was never even mentioned; almost certainly, the eugenic connotations were too strong. "If we cannot find a big enough figure for a dedication," Muller asked, "why have one?" Once again, his voice carried. They omitted the frontispiece.[38]

The first issue appeared in September 1949. The masthead was heavy with heredity clinic personnel: Nash Herndon from Wake Forest, Madge Macklin, now at Ohio State, and Norma Ford Walker from Toronto were all on the board. They were joined by Bronson Price, the New York physician and blood group expert Alexander Wiener, and Horace W. Norton. Neel did not appear, but Muller had his M.D.s.

Muller set the tone for the journal with an introduction that was, characteristically, overlong, combative, meticulously argued, and well substantiated, with a prescience that was almost a stylistic tic with him. He began by damning eugenics. In a section headed "Errors to Be Avoided," he wrote that something "flourished increasingly in the 'eugenics' of such racist propagandists as Lothrop Stoddard, Madison Grant, and Fritz Lenz. Finally, blossoming out into the Nazism of Hitler, it led to such excesses as to involve itself and a considerable portion of the world in ruin." The diplomatic Cotterman questioned the reference to Lenz, who not only was still alive but was a member of the society. He wrote to Muller, mentioning that he had discussed the essay with Lee Dice, and "raised the question as to whether you hadn't been perhaps 'a bit rough' on Fritz Lenz, and I wondered how he (Lenz) would 'take it.' I reminded Dr. Dice that we were at present exchanging reprints with Dr. Lenz. Dr. Dice quickly sat on me for being so timid, and remarked: 'as for the reprints, we'll exchange reprints with the Devil himself.'" Lenz, touchy about his Nazi past, fired off a defensive letter to Muller, but Muller stood his ground. (Lenz was not the only ex-Nazi struggling to return to professional good graces. Otmar von Verschuer was mounting a systematic campaign to rehabilitate his reputation, and Hans Nachtsheim was on the UNESCO committee on race.)[39]

Muller's essay continued by elaborating what became a standard scientific take on eugenics: it was bad because the science was bad; it got too tangled up with politics. It is ironic and oddly touching that one of the most politically engaged biologists in the country should strive so earnestly to distill science to an intellectual, apolitical essence. He articulated a remarkably flexible and subtle understanding of the relationship between genes and environment, explaining that it makes no sense to argue which is more important; this fact, however, by no means implies that hereditary improvement is moot, or that it is futile to try to disentangle hereditary from environmental effects. Looking to the future, Muller hoped someday to map the human germ plasm, as had been done with *Drosophila,* and he pointed to the little clusters of blood group genes, strewn along the human X chromosome, as evidence of progress. He hoped for many new markers along the chromosomes, but predicted that before that genetic knowledge came, there was "bound to be a great increase in cytological knowledge." Indeed, the minor discoveries were already being made that would lead, five years hence, to that great increase (chapter 6).[40]

Notwithstanding his criticism of eugenics as an obsolete social movement, Muller closed the inaugural essay in human genetics' flagship journal with a plea for cautious, compassionate, scientific eugenics. He defined eugenics in a Galtonian way, as "the social direction of human evolution." He called it a profound and important subject that would, in due time, be worked on seriously. It needed to be done, he wrote, carefully, calmly, humanely, and above all scientifically—without racial or class bias or undue emphasis on superficial traits. Much more knowledge needed to be gathered; much more basic research was the answer. Muller left no doubt in 1950 that great strides toward a science of human perfection had been made, and he saw the direction for the future.[41]

Cotterman backed Muller's essay with Neel's paper from the first ASHG meeting, on hereditary carriers of disease, and filled out the issue with articles that were heavily medical in orientation. C. W. Rucker contributed an article on color-blindness and sex-linked nystagmus, a form of involuntary eye movement. Clarence Oliver wrote an article on the genetics of tooth deficiency, a topic that Claude Nash Herndon had once laughed at as typical of egghead Ph.D. geneticists. Herndon himself contributed an article on hereditary muscular dystrophy derived from the Appalachian studies. Cotterman contributed an article on a new blood group, A3B, and there were articles on dominant hereditary ataxia and hereditary angioneurotic edemi. The early ASHG community was cozy, more medical than Muller might have liked, and firmly anchored in the heredity clinics of the 1940s.

The afterglow of putting out the first issue was short-lived. The $2,400 from the Eugenics Research Association had been only about half of what they needed to get the journal on a secure footing. Further infusions would be needed fast. Neel was, in principle, charged with raising funds for the journal, but, Strandskov wondered aloud to Muller, perhaps Franz Kallmann would be a more effective choice? Kallmann was a German expatriate and psychiatrist who had relocated to New York, where from 1937 to 1965 he ran the first psychiatric genetics clinic in the United States, at the New York State Psychiatric Hospital—the descendant of the Pathological Institute directed by Adolf Meyer and ancestor of today's Columbia University Department of Psychiatry. He made real contributions to psychiatric genetics, most notably a 1946 paper that established a genetic role in schizophrenia. He also hove resolutely to his belief in the importance of eugenics; every year from 1949 to 1964, he wrote a review article on psychiatric genetics called "Heredity and Eugenics." He was added to the board of directors following the 1949 meeting. It was apparently Strandskov who lined up the Coolidge Foundation to support the journal in year two; they came through with $4,000. For the moment, calamity was averted.[42]

Meanwhile, Cotterman himself was having trouble. Never emotionally very stable, in July 1950, "based wholly on a personal problem," he resigned both from the editorship of the journal and from his faculty position at Michigan, effective in October. That fall, he took a hiatus from the journal and traveled west, ending up at the University of California campus in Davis, near Sacramento. Even those close to him never understood exactly why he left, but as the department secretary put it, "it seemed to us he wanted to be free from contacts with various and sundry patients and able to devote himself entirely to problems which interested him without having to spend time on problems thrust upon him." In 1951 he returned to Michigan for a few months, then took an enormous backlog of manuscripts and holed up at the YMCA in his hometown of Dayton, Ohio, to finish volume 3 of the journal. "With another 5 or 6 weeks of solitary confinement in my YMCA cell," he wrote to Lee Dice at the end of June, "I think I'll be pretty well caught up. I preferred to work here this summer because of (1) the confinement, (2) the cheaper room ($6.75 per week), and the better restaurants, etc." When he finished, he headed back to Davis, where by 1952 he had obtained a position in the veterinary school.[43]

Then, disaster: in March 1951 the Coolidge people pulled their funding, despite having intimated that they would offer a second year's support. They were shifting their funding priorities to "the training of young men who are interested in the welfare of mankind in southeast Asia"—an area of vital importance as the Korean War heated up. The Cold War giveth and the Cold War taketh away. "This news presents a critical situation," Strandskov told the directors. "We can't operate without some outside help unless we curtail our activities." They were forced to consider combining or dropping some issues of the journal until more support could be found.[44]

In May another angel was found. Again it was Secretary-Treasurer Strandskov, aided by Dice, who lined up the support. Strandskov was coy about the patron's identity at first. But eventually it became clear that the benefactor was Wickliffe Draper, the leading philanthropic supporter of eugenic causes in the twentieth century, through the vehicle of his Pioneer Fund. Draper's $4,000 gift bailed out the journal again and enabled it to get on its feet. Strandskov was so happy that he wanted to sing Draper's praises in the journal, but Draper knew his reputation and did not want to bring ignominy to the society. They thanked him discreetly, by making him a "life member" in the society. Only the inner circle of the society knew that the nation's most rabid eugenicist had saved the day.[45]

Such gestures enabled Draper to keep a finger on the pulse of human genetics as it professionalized. He contacted many of the directors of medical or human genetic enterprises with offers to endow chairs or research projects. Most turned him down, but Nash Herndon and Wake Forest University found it in their hearts to accept the eugenic philanthropist's largesse. Herndon was a respected and well-liked member of the human genetics community. He had run the Wake Forest Department of Medical Genetics since Allan's death in 1943. He also headed the North Carolina eugenics society. Under his directorship, the rate of eugenic sterilization in North Carolina increased during the 1950s; most of those sterilizations were performed on poor whites. Herndon was an avowed eugenicist but not a white supremacist. Draper first funded Herndon's fieldwork, in 1950–51, at a level of about $100,000 per year. Then, in 1953–54, Herndon, Draper, and Wake Forest negotiated an endowed professorship. Draper stipulated only that the recipient not explicitly disavow his views, which included "To seek to have race and immigration laws maintained, enforced, and strengthened" and "To justify (explore) by scientific research the attitudes they reflect." Herndon was comfortable with that, and the donation went through in 1955. Aware of the inflammatory nature of his views, Draper insisted that the donation remain anonymous—perhaps they might call it the Harry Laughlin Chair, he suggested, or, failing that, the Charles Davenport Chair. The university and Herndon, however, were mindful of the reputation those men had, so in 1956 Herndon became simply the endowed "Professor of Medical Genetics." Thus, as professional human genetics emerged, the most medical of the American human genetics programs had the strongest ties to old-fashioned, racialized Progressive era–style eugenics. But an increasing sense of diplomacy masked those ties, as human-geneticists remade their image in a medical mold.[46]

The California banker and ardent eugenicist Charles M. Goethe took a different approach to backing the fledgling field. He preferred to sprinkle little gifts among the many, like some Dickensian benefactor tossing pennies to the poor. Graduate students in human genetics were among his favorites. For years, he sent Lee Dice small checks to buy Thanksgiving turkeys for Heredity Clinic students. And through the fifties, he sent checks to the ASHG treasurer, for gift memberships for students with high eugenic potential. "Gentlemen," he wrote to Eldon Gardner, the society secretary, in 1959,

> Whenever I can spare another $24 I plan to send you check for further memberships in ASHG with the hope you will use it for membership for someone struggling for a higher degree, particularly while he accepts the responsibility of fathering at least 3 children or say replacement.
>
> My own experience is that practically all of such not only are themselves of high IQ but that they have selected wives more companionable because of equal ability.[47]

The society cashed the checks.

Draper and Goethe understood the value of discretion; some lobbying groups were not so circumspect. In 1947 the organization Birthright, a eugenics group "exclusively devoted to promoting selective sterilization" based in Princeton, New Jersey, sought permission to print condensations of articles from the *Journal*. After the war, Birthright had helped lead a small boom in eugenic sterilizations, particularly in North Carolina, Virginia, and Georgia. Like most eugenics groups in the country, its members followed the activities of the society closely and regularly contacted human-geneticists. A couple of years later, the group asked Muller for permission to reprint his article "Genetic Prophylaxis." He agreed, but he refused their complimentary membership. "I am thoroughly in sympathy with the general aims of your Society," he admitted. Nevertheless, for political reasons, he could not risk public association. "I feel it very important, for winning the approval of groups who may well become dominant in the not distant future, that eugenics be so far as possible freed from the flavor acquired through its having been associated with Nazism." In response to the 1949 request, Herndon said, according to Cotterman, "I am sure that you know as well as I do that so-called eugenics organizations vary from the 'lunatic fringe' to some that are on a fairly sound basis. Birthright has long been one that has been well out on the fringe." The society refused Birthright's request to reprint the articles. The lines were not clearly drawn, however. Not long after, Birthright changed its name to the Human Betterment Association, a phrase with resonances of Kellogg's conferences decades earlier. In 1952, when Goethe offered to buy Lee Dice—then president of the ASHG—a membership in the Human Betterment Association, Dice thanked him but replied that he was already a member.[48]

The tension between scientific and medical approaches to human heredity suffused the rapidly professionalizing field of human genetics. Hermann Muller had been a godsend to the fledgling field. Selected before Hiroshima as spokesman and figurehead for the new society, he became one of the best-known American biologists with his Nobel Prize in 1946. For a few years, his passionate rhetoric and brilliant analyses placed human genetics firmly on the nation's front pages. In its first years, he was the face of human genetics. But the torch soon passed to Jim Neel. Where Muller relied on arguments of degeneration, Neel framed human improvement in medical terms that more surely set the direction of the profession.

In 1958 Neel published his own version of Muller's "Our Load of Mutations" argument. "Atomic Age culture," he wrote, not only was profoundly altering our environment but, most important, was accelerating the rate of change of that environment. Mutation rates and selective pressure were mounting to levels never before seen. At the same time, advances in medicine were diminishing our adaptability through humanitarian efforts. Modern medicine was in effect "thwarting the evolutionary mechanism for ridding the human species of the undesirable genes which are constantly arising." In other words, "at the very time when the species stands in urgent need of evolving to meet changing conditions, the medical profession may unknowingly be interfering with these necessary adjustments." Medicine, in short, was dysgenic—it was the same argument that had been made for decades. Few American scientists understood better than Neel the fundamental tensions in genetic medicine, the Garrodian versus the Galtonian. In forging a partnership between medicine and science, it constantly pitted the needs of the individual against those of the population. "In our concern for the individual," Neel asked, "have we forgotten to set up the team which has as its concern the species as a whole?" That was one point on which Muller would have vigorously nodded his assent.[49]

* * *

The atomic bomb was the best thing that had ever happened to human genetics. First, it released the war's inhibition on professional development. Herluf Strandskov had been trying to rally his heredity clinic colleagues into forming a society since before Pearl Harbor, but it simply wasn't viable until after the war. The sense of a new era having begun at the end of the war, then, is partly an artifact of the war itself having all but shut down the nascent field in 1941. But the bomb also provided a powerful new rationale for expansion and thereby accelerated the field's growth and professionalization. Anxiety over radiation drew professional geneticists back toward an interest in human heredity and particularly mutation. The bombs dropped on Hiroshima and Nagasaki provided a natural experiment that fueled years of research and debate and invigorated the field intellectually. And the fear of future bombs, perhaps dropped on America, suddenly made mutation a major health concern. Degeneration of the germ plasm, the bugaboo of generations of eugenicists, now had a frightening, real-world source. The founders of the American Society of Human Genetics used nuclear anxiety adroitly, to professionalize and garner funding, and particularly to shift the debate away from the human improvement that they longed for and toward the pragmatic, seemingly more attainable goal of preventing the unfortunate or undesired from being born. Up until now, enhancement had seemed the more benign eugenic strategy; elimination of the unfit had connotations of state control, bigotry, and even genocide. For the rest of the twentieth century, practices long associated with negative eugenics would be seen as the more benevolent, because they were medical, while enhancement acquired frightening connotations of building a master race.[50]

Negative eugenics had changed, of course, since the Progressive era. Most important, coercion had yielded to persuasion. Moving negative eugenics into the medical sphere allowed for a gentler, more volunteeristic negative eugenics. Persuasion was considered not just a right but a duty of the responsible physician; patients sought their doctor's advice actively and eagerly. But increasingly doctors and patients alike believed that the ultimate decision belonged to the individual—so long as the individual was sound of mind. The eugenic targets, too, had become more precise and specific: retinitis pigmentosa or peroneal atrophy, rather than blindness or crippling; amaurotic idiocy rather than feeblemindedness. Finally, the object of eugenic practice was shrinking. In the 1860s Francis Galton had described a plan for population improvement and the concomitant relief of suffering, because he believed populations to be more malleable than individuals. Immigration laws, too, operated at the level of populations. Sterilization and birth control can address human amelioration and improvement at either the level of populations or that of individuals, depending on how they are implemented. Both as medical and as eugenics practices, they permitted the shift from populations to individuals, making it unnecessary to worry about who gets married or moves next door. By the 1950s, physicians and scientists were beginning to conceive eugenics practice in terms of genes. From then on, molecular-level approaches to human amelioration—and betterment—would be the ultimate, if elusive, goal.

6

Getting Their Organ

"VICTOR MCKUSICK, 'FATHER OF MEDICAL GENETICS,' 1921–2008," ran the obituary headline from Johns Hopkins. The *British Medical Journal*, the *Lancet*, the NIH genome institute, and the March of Dimes all headed their obituaries the same way. The popular science magazine *Discover* referred to him as the "visionary researcher who is often called the father of medical genetics," and the online encyclopedia Wikipedia says that he is "widely regarded as the father of clinical medical genetics." "Rare is the scientist," wrote a *Science* magazine obituarist, "who is universally recognized as the founder of a field." Indeed, it is the sort of epithet that seems to place a scientist in the pantheon with Gregor Mendel, Charles Darwin, and Isaac Newton—people who lead revolutions, who create new disciplines out of whole cloth. The term *founder* does not refer to McKusick's prominent role in guiding medical genetics as it morphed into molecular genetics and genomics in the eighties or even the seventies; it refers to an origin myth of "medical genetics" from the late fifties through the sixties. McKusick's moniker refers to paternity, not parenting.[1]

It is a strange family where the child predates the father. When McKusick began studying heredity, medical genetics was not merely extant; it was, to those working in it, vibrant, with a history stretching back decades. In the Progressive era, physicians, public health workers, health reformers, and geneticists studied the genetics of disease, establishing research methods and disease model systems still in use today. The *Treasury of Human Inheritance*, the first systematic attempt to catalogue pathological heredity, dated to the first years of the century; surely Edward Blankenship, the English physician whose work with Karl Pearson formed the core of the *Treasury*, could have laid claim to parentage of medical genetics. In the twenties, Raymond Pearl had brought genetics onto the margins of the medical campus with his Division of Medical Genetics. Madge Macklin had coined the term *medical genetics* in print in 1932, and she and others had practiced it under that name since then. Laurence Snyder had taught the first course in human genetics for medical students in the country in 1933 and a few years later had midwifed the heredity clinics, where researchers approached many now-classic problems in medical genetics with the limited methods—prevention, including eugenic sterilization—and standard prejudices of the day. William Allan had founded the first Department of Medical Genetics in the country. In the late forties and early fifties, medical geneticists had gained a professional identity, with the creation of the American Society of Human Genetics and its flagship journal. Its founders had cast as wide a net as possible, embracing biologists, agricultural breeders, old-school eugenicists, anthropologists, and physicians. By the time McKusick became a father, then, his putative offspring already had a long but morally ambiguous history, in which the founding principles, methods, and administrative structures of the field were deeply entangled with some of history's worst abuses of biology.[2]

During the early years of McKusick's career, in the late fifties and sixties, genetics became established as a medical specialty, gaining standing in the formal and rather rigid hierarchy of American academic medicine. Scholars have called this a "critical period" and a "structural revolution," but a closer look at the preceding history reveals this moment to be a threshold, not a revolution. Medical schools had established genetics programs at a rate of about one or two every year since the founding of the heredity clinics in 1941. In 1956 Lee Dice retired, and as James Neel took the helm the Michigan Heredity Clinic became the University of Michigan's Department of Human Genetics. The next year three new programs began: at the University of Washington, the University of Wisconsin, and McKusick's at Johns Hopkins. McKusick's was the most ambitious and the most systematic. He colonized the Hopkins medical school, staking out administrative turf for his specialty and carefully delineating its boundaries and its internal structure. He systematized the knowledge of heredity and disease, through numerous publications, including a new catalogue of hereditary disease, the indispensable *Mendelian Inheritance in Man*. He proselytized for genetics as a medical specialty, through publications, conferences, and especially an annual summer course in Bar Harbor, Maine. And he baptized medical genetics in numerous synthetic and historical reviews that gave the field an internal narrative and creation myth. McKusick was neither visionary nor iconoclastic. His contributions are voluminous but modest, even low-status: clinical case studies, administration, teaching, cataloguing. He consolidated and codified genetic knowledge of disease in a clinical context, and thereby legitimated genetics among other physicians. He did not revolutionize medical genetics, but through his tireless institution building and professionalizing activities, he led the move to establish genetics as a specialty of mainstream medicine.[3]

Calling McKusick the father of medical genetics reflects an effort to impose a break on a smooth and troubling history. The study of human heredity and health evolved through the century, its methods, its professional identity, and its social relevance gradually accruing as it adapted to each cultural moment. A narrative in which McKusick is the father of medical genetics draws a comforting moral boundary between the contemporary genetic medicine we want to believe in as a positive moral good—and the earlier eugenics we want to sequester in the distant past.

Coincidentally, in the fifties and sixties, a number of technical advances, mostly small and low-tech, made academic hospitals attractive research sites for geneticists working on medical problems. I will examine biochemical and cell-culture techniques in detail in the next chapter, because their crescendo came later. More important for the legitimation of clinical genetics were the simple methods of studying chromosomes, long available to organismal geneticists and finally applied to man in the fifties and sixties. The unique identification of human chromosomes enabled medical geneticists to identify, for the first time, actual physical correlates of genetic disease. McKusick liked to say that other specialists all had an organ: the cardiologist had the heart, the nephrologist the kidney. Cytogenetics, he said, "gave us our organ." The germ theory of genes—the Galtonian, deterministic notion of gene as disease agent—became an organ theory of genes.[4]

* * *

Aptly, the alleged father of medical genetics was a twin. Victor A. and Vincent L. McKusick were born on October 21, 1921, on their parents' dairy farm in Parkman, Maine. Their father was a gentleman farmer—he was a college graduate and a former high school principal—and their mother had been a schoolteacher. Vincent studied law and eventually became chief justice of the Supreme Court of Maine, while Victor went into medicine. There was a "good environmental reason for that," McKusick said in an interview in 2001, playing on the nature-nurture dichotomy that is never far below the surface in human genetics. At age fifteen Victor developed a severe infection resulting from an abscess. Doctors treated him with the new drug sulfanilamide, one of the first commercially available antibacterial drugs. "In the process," he said, "I saw a tremendous amount of medicine and a tremendous amount of doctors and decided this was for me. Vincent, who did not have that experience, fortunately, continued his own course." Geneticists tend to like determinist jokes, and the writers among them savor the prickles of Latin syllables. "Perhaps I would have ended up a lawyer if it weren't for the microaerophilic streptococcus," McKusick quipped. He matured into a tall, rail-thin man, with an aristocratic air that aggregated his physician's authority, old-school formality, and reserved New England demeanor. Though he could be high-handed, he was kind and well liked. He had a knack for telling stories and a superb memory to the end of his life. He supplemented his memory with an ever-present camera, snapping candids at lectures and meetings like a parent at a playground. Part of his reputation as father of the field stems from his role as the field's first griot. His narratives are epic, precise, and uncannily consistent. They need to be handled with care.[5]

McKusick enrolled at Tufts University in 1940, but the war effort accelerated his plans. The medical school at Johns Hopkins, unable to fill its class, accepted McKusick after six semesters of college. Under accelerated wartime training programs, he earned an M.D. in three years, donning the white jacket in 1946. He did both internship and residency at Hopkins, and the university hired him to the faculty. He remained at Hopkins until his death in 2008.

McKusick went into genetics the way William Allan had done: he sought out a geneticist for informal tutelage. There was no other way. Hopkins did not yet offer a genetics course. His introduction to heredity came through the clinic: in 1947 a fourteen-year-old patient presented with melanin spots on his lips and inside his mouth as well as polyps in the small intestine. "Right after that, another single case came in," he said. "Then I had a family in which three members were affected, indicating that it was inherited." He heard "through the grapevine" that Harold Jeghers, a Washington, DC, physician, had five cases of the same combination of polyps and spots. He said he surveyed the literature and found that a Dutch physician named Peutz had described the syndrome in the 1920s. McKusick noticed the hereditary pattern, and, consulting Bentley Glass from the biology department, worked out the heredity. As Bateson tutored Garrod, as Snyder tutored Allan, so Glass guided McKusick through the elements of Mendelian genetics. McKusick published with Jeghers, although his name did not make it onto the eponym: Peutz-Jeghers syndrome.[6]

Polyps and spots did not turn McKusick into a medical geneticist. He went into cardiology and specialized in heart murmurs, utilizing the new technique of spectroscopy to analyze heart sounds. "I got training in cardiology because there was no such thing as medical genetics," he said in 2001. There was no such thing as medical genetics training at Johns Hopkins, but there certainly *was* such a thing as medical genetics. Indeed, the field was enjoying a growth spurt, triggered to a large extent by the recent formation of the American Society of Human Genetics. Preparing for his 1952 presidential address to the society, Lee Dice from Michigan polled his colleagues for news about new heredity clinics around the country. Beginning with the Cold Spring Harbor Eugenics Record Office, which he called the first heredity clinic, he recounted how the idea, which he had been so central in formulating, had spread to many institutions in the United States and Canada. The Michigan Heredity Clinic was thriving; Jim Neel was assuming increasing responsibilities and was the presumptive successor as Dice made preparations for retirement. Besides Herluf Strandskov at Chicago and Franz Kallmann in New York, there was the group in Utah, with F. E. Stephens and Eldon Gardner. At the University of Toronto, Norma Ford Walker was an associate professor of zoology and staff geneticist at the

6.1 Victor McKusick, on the right, with his twin brother, Vincent, c. 1924. Courtesy of Alan Mason Chesney Archives, Johns Hopkins University

Hospital for Sick Children. In Montreal, F. Clarke Fraser, an M.D. and Ph.D., had started up a division of medical genetics in the pediatrics department at McGill University's affiliated Children's Hospital, focusing on his specialty of dysmorphology, the study of skeletal and systemic defects. L. C. Dunn, Frederick Osborn's frequent adviser since the late 1930s, was planning an institute for the study of human genetic variation, which he expected to open in the fall. And, like a genetic Johnny Appleseed, Laurence Snyder seemed to leave behind a program in medical human genetics at every institution where he held a job or gave a lecture—among them Wake Forest, Ohio State, the University of Oklahoma, and soon the University of Hawaii. To those in this expanding circle, the growth of medical genetics was thwarted by the war, but it had resumed with added vigor in the late forties. In the early fifties, the golden age seemed just around the corner.[7]

McKusick was not yet part of that club. He had no formal training in genetics. He did not attend the AAAS or AIBS meetings where the ASHG members congregated. He was an adept networker, but his world was the parochial universe of the wards of Hopkins. In the course of his cardiological experience, he encountered Marfan syndrome, a condition involving heart problems, eye defects, and a tall, gangly stature (Abraham Lincoln is often suspected of having had Marfan). Like Dice and Herndon in rural North Carolina, McKusick sometimes spoke like a naturalist, collecting and doing descriptive taxonomy. "I collected a large number of Marfan patients," McKusick said, "and analyzed the families from the pattern of inheritance point of view and analyzed the individual cases from the point of view of the clinical manifestations and natural history of the disorder." He interpreted Marfan as a pleiotropic disorder—one gene, many effects. Like Garrod before him, he sought other similar conditions—in this case other heritable disorders of connective tissue. In 1956 he published a monograph under that title, which collected his findings on Marfan, Hurler syndrome, Ehlers-Danlos syndrome, osteogenesis imperfecta, and pseudoxanthoma elasticum, a peculiar condition in which the skin becomes so stretchy that it can be pulled several inches away from the body. The book established McKusick in the field of clinical genetics.[8]

* * *

That summer, he brought a talk on the heritable disorders of connective tissue to Copenhagen, where the Danish medical geneticist Tage Kemp presided over the first international congress of human genetics. The 1956 Copenhagen meeting looms large in the mythology of medical genetics, because it occurred at what has become known as the birth of the field.

McKusick called it a "very defining experience" in his medical-genetic education. It is a trough in the medical-genetic landscape; events on either side tend to roll mnemonically into the summer of 1956, and Copenhagen gets credit for publicizing and therefore originating them.[9]

Nearly four hundred delegates attended, and fourteen countries sent national committees to the meeting. The prewar and immediate postwar cohort who had established the heredity clinics and founded the ASHG still dominated the American committee, which included such representatives as Sheldon Reed, Pete Oliver, Eldon Gardner, and Arthur Steinberg. An ambiguous group of respected geneticists with Nazi ties represented Germany: Otmar von Verschuer, Fritz Lenz, and Hans Nachtsheim all sought reintegration into the international genetics community. The British committee included Lionel Penrose, the Galton Professor, and Harry Harris, a biochemical geneticist at London Hospital Medical College, among others. Many Scandinavian medical geneticists attended, of course. Their noncoercive medical eugenics established a model for volunteeristic state control of heredity; several delegates reported on recent efforts to institute genetic registration of infants as an experiment in socialized genetic medicine. The formal government apparatus provided a supporting structure for voluntary eugenics, which scientists such as Kemp deemed not only acceptable but necessary for the responsible stewardship of the race.

The Danish minister of education, Julius Bomholt, opened the proceedings by invoking the atomic age and how it had shaped human genetics as a field. He stressed its role in making the prevention of "the occurrence and spread" of hereditary disease a topic of intense interest. Kemp took up this question in his presidential address, sharpening it in genetic terms. Invoking H. J. Muller's paradigm-generating paper of 1950, Kemp wrote, "Within recent years, very much attention has been drawn to the dangers which our load of mutations involves for the human race." Indeed, Muller's paper, "Further Reflections on the Load of Mutations in Man," followed Kemp's remarks in the proceedings. But one could turn this observation around, Kemp noted, and recognize the "treasure of normal genes" we harbor in our cells. Medical geneticists were the stewards of the gene pool. "It is the task and responsibility of mankind in our generation, and in particular of the students of human and medical genetics, to protect this treasure and to shelter this heritage from harmful influences and threatening hazards." For Kemp, the "rise and rapid progress" of human genetics during the previous half-century—particularly in blood group studies, radiation genetics, population genetics, genetic epidemiology and control, medicogenetic registration, and genetic counseling—meant that the dream of genetic control was at hand. "The time is drawing near," he wrote, "when man can control his own biological evolution and also command his environments and conditions of life to an increasing extent." The medicalization of human genetics would enable mankind to at last realize the fantasy of self-directed evolution.[10]

Much of the meeting concerned topics that would have been familiar to any human-geneticist back to the beginning of the century. It featured twin studies and pedigrees; studies of inbreeding and cousin marriage; studies of color-blindness, hemophilia, and polydactylism; studies of race mixing; studies of intelligence and psychological disorders. Medicine, anthropology, and psychology still vied for predominance. President Kemp, describing Denmark's "medico-genetic or genetic-hygienic registration," outlined a method of field work indistinguishable from that of William Allan and Nash Herndon:

> A physician, who is trained as a specialist in the field concerned, makes a thorough investigation of the individuals with the disease or lesion in question and of their families, partly on the basis of hospital records, other documentary material and genealogical investigations, partly by traveling about, visiting and examining the individual patients in their homes or by calling the patients to an institution or to some hospitals for observation and more thorough investigation. Through the studies of the various diseases and lesions their mode of inheritance, their etiology and pathogenesis, their clinical picture, their frequency and geographical or social distribution in the population, the possibilities of their treatment and prevention, and the effective fertility of the affected can be investigated.[11]

Although analytical techniques had grown more sophisticated since 1940, the end was the same: "Using the experiences gained in the medico-genetic registry it will be possible to exercise a genetic-hygienic or eugenic activity as adviser on questions of sterilization, induced abortion, marriage, adoption and special relief." Such was the reality of preventive medicine in the atomic age. Although the bomb had rung in a new era filled with newly powerful sources of genetic risk, the means for reducing and ameliorating that risk remained about what they had been in the Progressive era. The gentle socialism of northern Europe provided the centralization necessary to consider such a project; American medical geneticists could only sigh and hope for such a system.[12]

New methods lay on the horizon. Until this point, prediction of genetic disease had been a statistical process: given known family history and the inheritance pattern of a disease, what are the odds that this couple's children will have a given condition? Two local physicians, however, were experimenting with a technique that could potentially tell whether *this* baby would have *that* disease. Earlier in the year, Povel Riis of Copenhagen County Hospital and Fritz Fuchs of the University Hospital of Copenhagen had published a letter to *Nature* describing a new method of determining the sex of a fetus. In 1949 the Canadian Murray Barr had discovered a dense, dark-staining body made of chromatin present in the cell nucleus of females but not males. A dozen years hence, the reclusive English cytogeneticist Mary Lyon would hypothesize (correctly) that the Barr body was an inactivated X chromosome. But for Barr, the tiny cellular structure was simply a clinical sign, an indicator of femaleness. From a blood test alone, he could establish a person's sex with near 100 percent accuracy: females were said to be chromatin-positive; males chromatin-negative. Carefully inserting a six-inch needle into a pregnant woman's belly, Riis and Fuchs successfully sampled the amniotic fluid, examined the cells, and ascertained the sex of the fetus. The same year, three other groups claimed to have discovered how to determine fetal sex, but again, Copenhagen gets the credit.[13]

In a pair of papers in the Copenhagen conference, Riis, Fuchs, and colleagues described the possibilities for extending the method. First, they used the "smear" method developed by Georgios Papanicolaou to determine the chromosomal sex of patients with syndromes of sexual development, such as Klinefelter's and Turner's. (The Klinefelter's patients were chromatin-positive; Turner's chromatin-negative.) Then they considered the possibilities for testing for genetic disease in the unborn. At this point, that was limited to serology. If the fetal blood group could be determined as early as the sex, they wrote—that is, by the fourth month—and the amniotic fluid could be safely withdrawn at that stage, "then it should be possible to diagnose both sex-linked and blood-group-linked hereditary diseases at a stage where pregnancy can be safely interrupted." The paper was as speculative as this makes it sound—they had only a couple of samples and were mainly describing the possibility. It was a method in search of a technique. But if it were successful, the authors saw potential applications in genetic preventive medicine. The determination of fetal sex and fetal blood group—and potentially other genetic properties in the future—they concluded, "would seem to be of value in preventive eugenics."[14]

Four years later, Riis and Fuchs reported aborting male fetuses from two mothers with hemophilia. Each boy would have had a 50 percent chance of having the disease. Under the Danish eugenics law, mothers had the right to request termination of pregnancy if there were "close risk that the child, due to inherited characteristics or to disturbance or disease acquired during foetal life, may come to suffer from mental disease or deficiency, epilepsy, or severe and non-curable abnormality or physical disease." What they called preventive eugenics we today would call prenatal diagnosis with therapeutic abortion.[15]

Such a procedure was still science fiction in 1956. Outside of X-linked traits such as hemophilia, which can be identified from a pedigree, there was still no significant cytogenetics of humans. Since the teens, *Drosophila* geneticists had been mapping traits to specific chromosomes. They had twenty years' worth of fine-structure mapping data, based on precise and consistent banding patterns brought out by staining the remarkable giant chromosomes of the fly's salivary glands. The genetics of other organisms—mice, maize, the bread mold *Neurospora*—had followed. But human genetics still relied on indirect methods. Although the debate over the correct number of chromosomes in human cells had been settled back in the 1920s, still the individual chromosomes could not be distinguished uniquely under the microscope.

Within weeks of Riis and Fuchs's *Nature* paper, however, a sheepish note appeared from Joe Hin Tjio, then working in Zaragoza, Spain, and Albert Levan, of the Institute of Genetics in Lund, Sweden: we seem to have had the chromosome count wrong. They had looked only at lung cells, and they noted that the result could conceivably be an anomaly, but their data strongly suggested that humans actually have forty-six chromosomes, not forty-eight.

McKusick remembered that at Copenhagen, "Tjio had an exhibit showing his chromosome spreads, which you had to count as forty-six, without a doubt, in number." Further, Charles E. Ford and John Hamerton from Harwell presented confirmatory evidence from a range of tissues. Copenhagen thus tends to be remembered as the place and time that they got the number right.[16]

The human chromosome number is difficult to determine under the best of circumstances—as Barker pointed out in 1927, the chromosomes are small and numerous, and tend to clump together in the nucleus. Further, the best techniques of the time for obtaining clear images of cell nuclei involved fixing, embedding, and slicing the tissue very thin—thinner than a cell nucleus—for viewing under a microscope. A given chromosome often spanned more than one slice. Integrating a tangle of chromosomes at different stages of condensation across multiple histological slices was fraught with inaccuracy. Four technical advances led to a new means of visualizing the nucleus. The first, tissue culture, was a significant and complex development that occurred over many years. Animal cells had been cultured for decades, but until 1950, human cells had proven refractory to culture techniques (see chapter 7). The discovery of colchicine, a toxic compound isolated from the autumn crocus, enabled researchers to suspend the cell cycle at metaphase, the stage in which the chromosomes were most countable, looking like little sausage links. The other developments were humble, almost trivial. In 1952 T. C. Hsu at the University of Texas accidentally added deionized water rather than saline to his culture medium. The cells took up the pure H_2O and swelled, spreading the chromosomes out within the nucleus. And an old, low-tech method of preparing chromosome slides that the plant biologist John Belling had developed in the twenties was at last applied to preps of human cells: rather than fixing, embedding, and slicing, you simply put a drop of culture medium onto a microscope slide, added a coverslip, and squashed it flat with your thumb. This leaves the chromosomes intact, removing the ambiguity of integrating chromosome counts across slices. All of these techniques had been available for years. Tjio and Levan appear to have been the first human-geneticists to combine all four.[17]

Once researchers recalibrated their expectations, they were flooded with data. Cytogeneticists trawled the margins of society for chromosomal anomalies among primitives, criminals, the mentally ill, and the sexually ambiguous. At the radiation genetics laboratory in Edinburgh, where Muller had brought his X-ray machine back in the 1930s, a mutation group led by William Court Brown, A. G. Baikie, and Patricia Jacobs examined sex chromosome anomalies, as did Paul Polani's group in London. They linked "chromatin-positive" Klinefelter syndrome to the presence of an extra X chromosome, leading to an XXY constitution, while "chromatin-negative" Turner syndrome correlated with the loss of an X. They found a "super female" who had a triple-X constitution. And in 1962 they found an XYY "super male" in a population of "defectives" in a nearby mental hospital. Court Brown speculated that there might be a correlation between violent crime and the extra male chromosome. The bad blood, bad germ plasm, and bad genes of earlier days had given way to the bad chromosome.[18]

Soon, researchers also linked non-sex chromosomes, or autosomes, to disease. In 1959 the French physician Jerome Lejeune found that "Mongolism"—which Penrose was campaigning to rename Down's (today simply Down) syndrome—correlated with an extra copy of the smallest chromosome. Presence of an extra chromosome is called a trisomy. Most trisomies are lethal in humans; they result in miscarriages. But several are viable, and within five years, researchers had identified all of them. Although the chromosomes could be counted, it was still several years before they could be uniquely distinguished. Well into the sixties, recalled the cancer geneticist Peter Nowell in 2007, "you couldn't tell a 21 from a 22, or a 7 from an 8." Absent unique identifiers, researchers grouped similar-sized chromosomes into lettered clusters, A through G. Alternative nomenclatures proliferated. Like the gene nomenclature debate in the 1940s, the seemingly desiccated topic of chromosome standardization and nomenclature in the sixties became partisan and passionate. In 1960 Theodore Puck in Denver organized a workshop to negotiate a consensus, but it did little to cool researchers' tempers. Too many investigators had invested too much to make any one naming system universally acceptable. "The risk that a minority may be unable to accept the system as a whole," cautioned the workshop's summary report, "should not be allowed to delay adoption by a majority." They agreed to number the chromosomes in descending order of size, starting with the largest, designated chromosome 1. But lingering ambiguities in cytological technique resulted in the smallest two being swapped; Down syndrome, trisomy of the smallest chromosome, is trisomy 21. Uncertainty remained throughout the decade. Even in 1969, McKusick was still referring to "trisomy D" and "trisomy E."[19]

Biochemical genetics, still on the edge of the radar screen, represented an alternative to the organ theory of genes. The origins of biochemical genetics lay in blood group serology. Blood group antigens are excellent genetic traits; everyone expresses them and there is no ambiguity over which form someone has. In recent years, researchers developed new methods for analyzing and purifying other proteins as well (see chapter 7). Given a set of clinical symptoms, phlebotomy and a few laboratory tests could identify blood titers of molecules and compounds that correlated with disease. Combining this method with pedigree analysis, twin studies, and/or consanguinity studies, the researcher might be able to make strong claims about the inheritance of the trait. The researcher still did not "have the gene," but with luck and skill one might get what appeared to be the gene product. Within a few years, biochemical markers began to provide reliable signs and predictors of simple Mendelian disease.

In 1961 the microbiologist Robert Guthrie developed a test for the rare amino acid deficiency phenylketonuria (PKU). His method was a simple bacterial inhibition assay: bacteria are cultured on an agar plate with a compound (B-2-thienylalanine) that inhibits growth. Phenylalanine, the enzyme missing in PKU patients, overcomes the inhibition. Thus blood from a normal patient placed on the agar produces a bacterial colony, while PKU blood does not. With modest training and skill, doctors and nurses could learn to draw blood and fill the standardized spots on a "Guthrie card" and send it to a lab where the culturing was performed. In 1962 W. J. Culley, working in Canada, adapted paper chromatography to develop a more involved but more quantitative measure of phenylalanine production. By middecade, both the United States and Canada had implemented mass screening of newborns. In the seventies, expanded genetic screening encompassing wider geographical areas and more common diseases would lead to the eruption of major ethical debates over medical genetics and fuel the growth of professional genetic counseling. Archibald Garrod's inborn errors of metabolism, conceived originally as nonpathological markers of individuality, had come to signify the agents of disease. Garrod had been interested in "soil," but for most researchers, his inborn errors now connoted "seed."[20]

Yet Garrod's original sense of the inborn error had begun a small renaissance. The 1956 Copenhagen meeting contained a small session on biochemical genetics; almost every paper in it invoked Garrod, either directly or by referencing the term or concept of inborn errors of metabolism, often in the context of constitutional "soil." The neoconstitutionalist Roger J. Williams led off, making the case for his "genetotrophic concept," an updated, biochemical take on George Draper's "panels" of constitution (chapter 3). In 1956 Williams published *Biochemical Individuality*, a semi-popular monograph describing the genetotrophic concept and of course invoking Garrod in the title. Williams distanced himself from eugenics, which he thought impossible to carry out with justice, and articulated the concept that "heredity plays a role in *all* diseases; infectious, nutritional, metabolic, degenerative, mental, psychosomatic."[21]

The notion that heredity plays a role in all disease implies that genetics is the fundamental science of medicine—that it underlies all biological processes and therefore is involved in every aspect of health. A biochemical approach to genetics makes such a view possible (though by no means necessary) thus: genes were coming to be understood as somehow specifying proteins. Proteins do the work of the cell; they constitute the receptors, enzymes, signals, and structural molecules without which we would be inanimate bags of fat, sugar, and water. If disease is a response to the body's environment—and who could argue that it is not?—it is somehow mediated by proteins. And therefore by genes. If a given disease is a maladaptation, a genetic variant may underlie it. If a disease is a normal response to a toxin or germ, there may be genetic variants that confer resistance. In 1902 Garrod had written of the genetic basis of minor chemical variations in metabolism, obesity, and the "various tints of hair, skin, and eyes," as well as "idiosyncrasies as regards drugs and the various degrees of natural immunity against infections." By the mid-1950s, a small contingent of medical geneticists had begun to realize Garrod's vision.[22]

In practical terms of building a nascent discipline, this all-encompassing approach implies a strategy of infiltration. If you believe genetics lies at the foundation of all medicine, and you want to spread the influence of genetics, what you need to do is inject a bit of genetics into all the other departments and divisions of the hospital. Consult wherever you can, show your colleagues how an understanding of genetics will improve their practice.

McKusick's colleague Barton Childs was an infiltrator. His approach to medical genetics provides a nice contrast to McKusick's. Born on Leap Day 1916, Childs grew up in Chicago; till the end of his life he was a flat-voweled midwesterner—candid, cantankerous, and deeply loved by those who knew him. He was adopted, which amused him as a geneticist, he said, "because I have no family history." He was fiercely private: he refused to discuss his personal life or feelings in interviews, and disliked being photographed. Childs had a scientific temperament—analytical, experimental, quantitative. Like McKusick, he came to Hopkins for medical school, joined the faculty, never left. He became interested in pediatrics and joined the Harriet Lane Home, Hopkins's pediatric center.

Because hereditary conditions are present from birth, pediatricians see a disproportionate amount of genetic disease. "There was a tremendous number of children with anomalies," he said in a 2001 interview, "and I wondered what was known about them and read something about anomalies and learned that there were two ways to study them. One was to take something out of every bottle on the shelf and give it to a pregnant rat, and not surprisingly, the rat would have deformed offspring."[23] This was the clinical specialty of teratology. F. Clarke Fraser at McGill University was a pioneer in this approach. "Everything was going towards environmental causes of malformations," Fraser said about his own entry into medical genetics, in the late 1940s, "and I thought you ought to get genetics back into the picture." A plastic surgeon named Happy Baxter, he said, suggested that he give the newly discovered steroid hormone cortisone to his experimental animals, hypothesizing that he would see defects in the developing spinal cord, or neural tube. "So," Fraser said,

> I stuck some into some pregnant mice, making wild guesses as to what the dosage was and so forth. And we got cleft palates, not neural tube defects. We showed very early that there were mouse strain differences in frequency from the same dose at the same stage and everything. That was the first demonstration, I think. It was the first normally used drug that would cause malformations in mice. It also was the beginning of bringing genetics into teratology.[24]

6.2 Barton Childs, the Garrodian with no family history. Courtesy of Alan Mason Chesney Archives, Johns Hopkins University

"That seemed a rather inelegant way of doing things," Childs said. "I found that the alternative way was to do genetics." It is typical of the rivalries between subspecialties to deny that the other is even doing genetics. "The genes seemed circumscribed and rather fine in how they worked, and that seemed a far superior way to understand the production of anomalies." Childs grew into one of the most passionate and articulate Garrodian medical geneticists—clinical, biochemical, individual—and inspired many contemporaries and younger medical researchers to follow his path.[25]

As two of the few doctors at Hopkins with an active interest in genetics, Childs and McKusick connected. With Bentley Glass from Biology and Abraham Lilienfeld from Public Health, they formed the Galton-Garrod Society reading group (chapter 1). They also developed lectures on genetics that they presented to the medical students, although after a few years McKusick took over the course himself. Although Childs and McKusick established medical genetics at Hopkins together, Childs gets less acknowledgment, even at his home institution. Childs, steeped in the Galton Institute's brand of Garrodian individualism, conceived of genetics as something fundamental to all of medicine. This meant trying to make the various medical specialties more genetic, rather than establishing genetics as its own specialty. His style of intellectual infiltration eventually colored all of Hopkins' biomedicine. Today, Childs's philosophy of medicine, expressed in his dense, elegant book *Genetic Medicine: A Logic of Disease*, lies at the core of the Johns Hopkins medical curriculum Genes to Society, but he never became the "father" of anything professionally.[26]

* * *

Whereas the Garrodian notion of genes-as-soil implies a strategy of infiltration, the Galtonian genes (or chromosomes)-as-seeds approach implies colonization. McKusick proved to be a master colonizer. He had a receptive intellectual and professional climate, unmatched resources and institutional reputation, and the entrepreneurial acumen to exploit them in building an institute of international renown. He used them to build at Hopkins the most successful heredity clinic in the country, nestled in the bosom of one of the hospital's flagship departments.

Hopkins was known as the American bastion of "scientific medicine." Traditionally, that had meant science as the handmaiden of medicine, the rational explication and analysis of disease. In the Progressive era, the constitutionalist and eugenicist Lewellys F. Barker established clinical laboratories within the Department of Medicine, several of which gained administrative standing as "divisions," or subdepartments. Under this structure, Hopkins became increasingly supportive of curiosity-driven basic research and known for the study of rare conditions. In 1929 Joseph Earle Moore took over the Hopkins syphilis clinic, established in 1915. It epitomized Barker's vision: in it, a clinic and laboratory existed side by side, one treating the disease, the other studying its biology. With the introduction of antibiotics after the Second World War, syphilis became much less interesting therapeutically; it seemed basically solved. Moore expanded the syphilis clinic into a generalized chronic disease clinic. He established ties with researchers and clinicians around the medical campus—such as epidemiologists in the School of Hygiene—and at other universities. He developed an informal exchange program with several hospitals in the United Kingdom, with fellows traversing the Atlantic in both directions. He started a new journal, the *Journal of Chronic Diseases,* that fit his broader interests. Beginning in July 1955, Moore published in serial form many of the chapters of McKusick's *Hereditary Diseases of Connective Tissue.* Thus the study of infectious disease morphed into the study of genetic disease.[27]

In 1954 Moore was diagnosed with prostate cancer; he would retire in 1957. His clinic would need a new chief. A. McGehee Harvey, chairman of the Department of Medicine, had been expanding Barker's system of specialty divisions. He tapped McKusick as the next head of Moore's clinic. McKusick easily negotiated the establishment of a new Division of Medical Genetics, which would be coextensive with the clinic but administratively part of the hospital chain of command. The division of medical genetics was formalized on July 1, 1957. Deftly, McKusick renamed Moore's clinic the Joseph Earle Moore Clinic.[28]

With the clinic, McKusick inherited Moore's patient population and referrals from around the hospital. He assumed Moore's grants and maintained his ties to the United Kingdom. He inherited Moore's tradition of hosting English fellows, who suddenly found themselves being trained as medical geneticists. “Many of them came early on to work in a chronic disease unit,” McKusick said, but found themselves working on genetic problems. “I really proselytized them to the field of medical genetics.” For example, Edmund Anthony (“Tony”) Murphy was already a fellow in the Moore Clinic when McKusick took it over and steered him toward human population genetics. Samuel H. (“Ned”) Boyer IV came to Hopkins to work on heart sounds; McKusick pushed him toward genetics as well, partly through sending him back to England for a stint at the Galton Laboratory to learn biochemical genetics.[29]

McKusick himself rounded out the section by running “clinical genetics”—genetic counseling and diagnosis—thereby setting clinical and scientific medical genetics side by side. One of the first projects he undertook was to follow up on the patients from the largest study of constitutional medicine to have been done at Hopkins: Raymond Pearl's longevity study from the 1920s. Pearl's study, recall, was a clinical updating of the methods of Davenport and the ERO (see chapter 3). McKusick recognized the value of Pearl's data and realized that he could multiply that value by following up with the many patients who were still alive, as well as their children. He gained access to Pearl's files and even hired Pearl's assistant, Blanche Pooler, out of retirement to help him. He made the longevity study longitudinal by tracking down patients and death records and even using “field work sheets”—direct descendants of the record forms employed by Pearl, Davenport, and Galton. Another nod to earlier methods of medical genetics was McKusick's study of the genetic diseases of the Amish, a set of reproductively isolated communities in Maryland, Ohio, and Indiana, which in some ways were an even better group than William Allan's “gold mine of heredity” in the North Carolina mountains.[30]

In February 1959 McKusick brought in Malcolm Ferguson-Smith, another English postdoc, to set up a human cytogenetics laboratory—among the first in the United States, if not the first. Back in the teens, when Thomas Hunt Morgan's *Drosophila* group developed its gene-mapping techniques, the publications practically flew out of the labs. So it was for human cytogenetics in the sixties. Although proficiency required practice and a virtuoso cytogeneticist was a master indeed, the work required little specialized or expensive equipment, and postdocs, graduate students, and technicians could be easily and quickly trained to begin getting results. The data, once standardized, were immediately interpretable and were directly relevant to clinical diagnosis. In the 1960s, cytogenetics had a kind of glamor; it had the same sex appeal as the germ theory in the 1880s or eugenics in the 1910s—the allure of finding a single causal agent for a medically important condition.[31]

Ferguson-Smith established "sort of an assembly line" of cytogenetics, recalled Barbara Migeon, who worked in the lab, then ran it when Ferguson-Smith left. Once a good-quality photograph was taken, in which all the chromosomes could be seen and measured, the image would be enlarged three thousand or four thousand times. Workers would then physically cut out the individual chromosomes and sort them—chromosome 1, chromosome 2, X, Y, and so forth. "We would all sit there and cut out chromosomes from photographs and label them and put them in envelopes," Migeon said. "Then eventually you'd paste them," she said, into an orderly display called a karyotype. The process would have been familiar to Barbara McClintock, working on maize chromosomes in the 1930s, or even Calvin Bridges and Alfred Sturtevant, working on *Drosophila* in the teens and twenties. Once the display was standardized, one could search for anomalies and correlate them with the medical condition of the donor. "We were looking at all kinds of individuals for the first time, and there were lots of hypotheses about who might have an abnormal chromosome," Migeon said.[32]

Sex chromosome anomalies and autosomal trisomies involved the presence or absence of an entire chromosome. Soon, cytogeneticists detected gross rearrangements, in which a piece of a chromosome was lost, duplicated, or "translocated" from one chromosome or another. The Ferguson-Smith lab specialized in the sex chromosomes and especially sex chromosome anomalies such as Turner and Klinefelter syndromes. They also described chromosome patterns in different human diseases and different animal species, and they looked at the chromosomes of what they still called "mental defectives."[33]

In 1960, ninety miles to the north, Peter Nowell, a physician-scientist at the University of Pennsylvania, and David Hungerford, a postdoctoral fellow at nearby Fox Chase, found what at first appeared to be a new, albeit tiny, chromosome. "We were looking at different kinds of leukemia" and trying to find a chromosomal correlation, Nowell told me. The British groups at Harwell, London, and Edinburgh had been doing similar work. In Philadelphia, the staff at nearby Presbyterian Hospital would draw blood, or allow Nowell to draw blood, from leukemic patients so that he and Hungerford could look at their chromosomes. In the acute leukemias, they found no consistent chromosome abnormality. In one of the chronic leukemias, however—chronic myelogenous leukemia, or CML—they found "this little abnormal chromosome that was present in every cell of every case."[34] They realized it had to be a chromosome fragment. "What we couldn't tell," Nowell said, "was whether it was a deletion or a translocation, because the piece that was missing was so small that you couldn't, in those days without banding, tell whether the missing piece was stuck on some other chromosome."

Drosophila geneticists had long been able to stain chromosomes so as to highlight characteristic patterns of light and dark stripes, which acted like signatures for the various parts of the chromosomes. In 1917 Alfred Sturtevant had described in *Drosophila* a rearrangement called an inversion—in which a chromosome segment breaks off and reattaches upside down—by observing a reversal in the banding pattern. But there was no way to make such discriminations with human chromosomes. All the human-geneticists had to go on in the sixties was overall shape and size. "It wasn't until the seventies," Nowell said, "when banding techniques came along and specific cytogenetic changes in other leukemias, associated with other leukemia types, were identified, that people came to really realize that this was a way of identifying a particular genetic change that was clearly central to the causation of the tumor." The "Philadelphia chromosome" was a landmark: the first cancer associated with a specific chromosomal change. But technically, it merely represented human cytogenetics slowly catching up to where Morgan's fly boys had been back in the teens.[35]

McKusick emerged as the leader of a largely descriptive science of medical genetics that was driven by the needs of the clinic. The character of his Division of Medical Genetics at Hopkins was colored strongly by local institutional traditions and style—and by McKusick's skill in following and using them. Genetics was always the handmaiden of medicine; McKusick's reference point was always disease—a set of morbid or debilitating symptoms. Like Maryland's fertile forests of mixed hardwood, Hopkins Hospital was rich in resources and had intense competition; it was teeming with organisms vying for light, space, and nourishment. In this environment, McKusick competed successfully for resources by staking hereditary disease as his territory.

Although in the long run, Childs's integrative, Garrodian approach permeated the medical school, in the sixties it was McKusick's colonizing strategy that grew the field and established his reputation. McKusick was interested in discipline building. When asked why he asked for a division rather than his own department, McKusick replied that "it would have been inconceivable" to ask for a department. As director of a tiny department, he would have been the runt of the litter, always the last to the teat. As a division chief, he gained access to the director of the Department of Medicine—with Surgery, one of the two most powerful men in the hospital. Childs, in contrast, had no interest in administration or politics. "I declined to set up a genetics clinic" in the department of Pediatrics, he said. He preferred to pollinate the other pediatric clinics with genetics. "I thought that I would be a resource for the department for people who had families with genetic diseases." Where Childs's ideas sit almost anonymously at the core of Hopkins medical education, McKusick's name is on the letterhead of Hopkins's McKusick-Nathans Institute of Genetic Medicine. While Childs developed a reputation as a visionary, McKusick became known as the father of medical genetics.

* * *

A different sort of ecology predominated in the Pacific Northwest, where Arno Motulsky built a medical genetics program at the same moment as McKusick. Like Wake Forest in 1940, the medical school at the University of Washington, in Seattle, featured little crowding or competition but scant resources. It was established in 1948, and five years later its programs were still filling out. Robert H. Williams, an endocrinologist who had spent time at Hopkins, Harvard, and Vanderbilt, was the chairman of medicine. He had limited money but ample freedom to offer as he selected his faculty. As McKusick brilliantly exploited the specific resources available to him at Hopkins, so Motulsky adroitly took advantage of Washington's penurious liberty to craft a new institute.

Born in 1923 to a middle-class Jewish family in eastern Germany, Arno G. Motulsky made a harrowing escape at age seventeen from Hitler's Germany; the story is chronicled in the acclaimed nine-hour documentary *Shoah*. Landing in Chicago, he said, "My whole life was now in front of me. I didn't have that many choices, because my father didn't have any money—I didn't have any money." Learning to get along without money became a specialty. He worked as an animal caretaker in the virus research laboratory of Michael Reese Hospital, where he learned about virology. Taking night classes thrice weekly, he amassed enough credits to apply for medical school at the University of Illinois, in Chicago. In 1943 he joined the U.S. Army. Like many of this second generation of medical geneticists, Motulsky benefited from the V-12 program, which provided for accelerated medical training. Finishing his B.S. degree concurrently with his first year in medical school, he earned an M.D. from Illinois in 1947.[36]

Choosing an internship back at Michael Reese Hospital, he did a laboratory rotation in hematology, where he became interested in sickle cell anemia. He met the geneticist Herluf Strandskov, one of the founders of the American Society of Human Genetics. He soon realized that blood and genetics went very well together. He loved research, and he cared about clinical work. He began to focus on anemias.

In 1953 he got a call from Robert Williams at Seattle. In assembling his medicine faculty, Williams sought talent rather than balance. When he called Motulsky, he had already hired the respected hematologist Clement Finch. Williams, Motulsky says, "figured two quite independent hematologists would be maybe too much and that I should have my own kind of program. So he very smartly, in 1957, offered me to start a unit, a division of medical genetics in the Department of Medicine, which was really unheard of." They had heard of it in Baltimore, Toronto, Salt Lake City, Minneapolis, Ann Arbor, Madison, Austin, Columbus, Wake Forest, Norman, Hamilton, and Saskatoon, but not many other places.[37]

"I had nothing to lose," Motulsky said. It was a new institution with little money and no reputation but a strong young adventurous faculty. "There was no tradition," he said, "so people could build up the way they thought it was best." This made the administration particularly open to experimentation. "If you had ideas and so on, they let you work at it," he said. He was able to carry out research as well as see patients, to find his own mix of clinical and research work. He started giving "bootleg lectures" in genetics under the rubric of hematology. Motulsky did not romanticize this as a time of standing up against the criticisms of an establishment; rather, his narrative is one of autonomy and pioneering. "Really, I didn't have encouragement nor discouragement. They said, 'Well, sounds like an interesting opportunity.' "[38]

Motulsky identified as an infiltrator and a proselytizer. He "spread the message of genetics," he said, through giving rounds (medical lectures) and other talks; these led to consultations, which increased visibility. He traveled Europe, visiting the medical genetics clinics in Scandinavia, France, and Great Britain. Like Barton Childs at Hopkins, he was particularly impressed with Penrose and Harry Harris in London. When he returned, he successfully petitioned Williams for a teammate. And like Lee Dice at Michigan, Motulsky sought to balance the clinical and basic research sides of his operation; he needed a Ph.D. on his team. He identified Stanley Gartler, a biochemical geneticist at L. C. Dunn's Institute for Human Variation at Columbia.

Gartler and Motulsky specialized in G6PD deficiency. Like sickle cell anemia and thalassemia, G6PD (glucose-6-phosphate dehydrogenase) deficiency is a biochemical-genetic response to malaria. Individuals with a single copy of the mutant G6PD gene are resistant to malaria; those unlucky individuals who get two copies of the mutation have a tendency to hemolytic (blood cell–rupturing) anemia. G6PD deficiency was first discovered in the 1920s, among workers on South American banana plantations run by the United Fruit Company. Bayer Pharmaceuticals was testing new antimalarial drugs and found that 10 percent of the black banana pickers developed strong anemia in response to the experimental drug primaquine, a forerunner of modern antimalarials such as chloroquine. G6PD deficiency went on to become a kind of model system for human biochemical genetics. In a 1957 article now considered a classic, Motulsky cited G6PD deficiency as an example of a new line of research that promised to uncover the genetic basis of idiosyncrasies in the patient's response to drugs. Other examples included a heritable resistance to the muscle relaxant succinylcholine and liver damage caused by quinine. Because G6PD deficiency was a crude genetic marker, it also gave a new impetus for race as a medically relevant trait: "Since a given gene may be more frequent in certain ethnic groups," Motulsky wrote, "any drug reaction that is more frequently observed in a given racial group, when other environmental variables are equal, will usually have a genetic basis. Investigations on drug reactions therefore should include careful notation of the ethnic or racial extraction of the patient." Motulsky's paper is often cited as a founding document in the field of pharmacogenetics, the study of genetic variation in response to drugs, and is considered a model example of Garrodian medical genetics. Indeed, Garrod, in his classic 1902 paper on alkaptonuria, had discussed species differences in response to drugs and infecting organisms, which, he said, presumably have a chemical basis (see chapter 1). Two years after Motulsky's paper, Friedrich Vogel of Heidelberg, one of the first of a new generation of German human-geneticists, coined the term *pharmacogenetik*. Vogel spent time with Neel in Ann Arbor and became friends with Motulsky, later publishing a textbook of human genetics with him.[39]

Motulsky, then, was cast more in the mold that made Barton Childs than that of McKusick. Intellectually, he is biochemically oriented, Garrodian, interested in human variation. He thinks of genetics as fundamental to all of biology. His administrative temperament was infiltrationist. But his career trajectory was very different from Childs's. The liberal, flexible environment in Seattle allowed him to homestead a new program at a new university, without much competition for resources or defense of territory. Thus he established the same type of unit there as McKusick had done at Hopkins, even though he had the opposite administrative style. As in nature, nurture matters.

* * *

Whereas for McKusick at Hopkins it would have been "inconceivable" to ask for a separate department, in Madison, Wisconsin, it would have been foolish not to. There, the brilliant young microbial geneticist Joshua Lederberg established a model for medical genetics as a basic science. Compared to the programs at Hopkins and Seattle, the Wisconsin department was the least clinically oriented. In some ways, this program is the best model for late-twentieth-century genetic biomedicine, because later programs did indeed "cross the street" from the clinical to the basic sciences.

In 1947 Lederberg was finishing up a virtuosic dissertation with the biochemist Edward Tatum, in which he demonstrated "sex" in bacteria. As he wrote up his work and presented it at meetings, he was flooded with job offers. The most attractive came from the University of Wisconsin at Madison. Wisconsin boasts the oldest department of genetics in the country. It was founded in 1910 in the university's agricultural college, as the Department of Experimental Breeding, a Progressive application of science to farming. In 1918 it changed its name to Department of Genetics, and over the following decades it evolved into a distinguished program in research genetics, particularly plant genetics. The applied nature of the genetics program fit well within the "Wisconsin model" of state-sponsored research that in turn benefited the state.[40]

"There was determined opposition to Lederberg's appointment," recalled the maize geneticist Royal Alexander Brink, the department's longtime chair. Brink drily noted that Lederberg, a New York Jew to whom New Haven seemed provincial, had a "metropolitan" background. But Lederberg was convinced that the faculty's reservations about his urban upbringing veiled their anti-Semitism, which indeed was widespread in American universities at this time. As he put it later, the department decided, "Lederberg's a Hebe but he's so damn smart let's take him anyway."[41]

Lederberg hardly knew a combine from a cow pie, and he didn't care to. His orientation had always been medical rather than agricultural. He had spent from 1942 to 1944 at the College of Physicians and Surgeons of Columbia University, before switching to basic science at Yale. He always identified with medicine, claiming that his early training gave him a breadth and perspective most scientists lack. Lederberg's growing interest in mutation, combined with an abiding passion for social responsibility and involvement in political issues relating to science, led him to human and medical genetics. In the late forties and early fifties, he supported his research with funds from the Atomic Energy Commission (AEC). At a dinner at the house of geneticist Curt Stern in Berkeley in 1953, he met John Zimmerman Bowers, a Maryland native and an M.D. who, as Deputy Director of the AEC's Biology and Medicine Division, had gone to Nagasaki in 1949 and later shielded his eyes from the test blasts at Eniwetok atoll. Since then, he had moved into medical education. At the dinner, Lederberg recalled "berating him for the absence of genetics in medical schools."[42]

In 1955 Bowers became the new dean of the School of Medicine at Wisconsin. In contrast to the nationally known school of agriculture, the medical school was solid but undistinguished, regional in orientation, and focused more on teaching than on research. Bowers set about with an ambitious program of reforms to transform it into a world-class medical school, including formalizing and streamlining the administration and beefing up the research programs. These activities did not endear Bowers to everyone—in 1961 he resigned amid scandal—but he and Lederberg got on well. In December 1955 Lederberg sent him a memorandum outlining his ideas for a program in medical genetics. It would emphasize collaboration with other medical school colleagues, research, and teaching (there was no mention of clinical work). It could take one of three forms: an informal working group, comprising faculty from other departments; a division, or subdepartment, within a large department such as medicine or pediatrics; or an independent, stand-alone department. The first two options had problems, Lederberg said, the chief one being that without exceptional support from the department chair and faculty, the program might "wither on the vine." A separate department brought administrative duties, yet Lederberg wondered coyly and parenthetically "how strenuous could they be in a 'one-gun department'?" At the working end of the barrel, of course, was Lederberg. In recommending a separate department, he was writing himself into an aspiring administrator's dream job description: a chairmanship with few or no other faculty.[43]

Lederberg expressed an original, if idiosyncratic, vision of medical genetics. "Medical genetics has been considered synonymous with human genetics," he wrote in the 1955 memo. Not synonymous with but subordinate to: recall the debates over the house organ for the American Society of Human Genetics in the forties (chapter 5). The founders thought of disease states as a portion of human existence. But for Lederberg, health was the general rubric, with human biology just one part of the equation. In his mind, medical genetics therefore subsumed human genetics. Lederberg identified his medical training as broadening, in contrast to what he saw as the narrowness of science. "I started out as a medical student," he told me in 1996. "That's the broadest possible thing. Highly interdisciplinary: you've got microbiology, biochemistry, pharmacology, physiology, anatomy, and how this relates to pathology. You're looking at disease changes and the relationship of disease to the social milieu. Clinical observation is natural history." The genetics of disease-causing bacteria and viruses counted as medical genetics, as did the genetics of model organisms for human disease. To Lederberg, the medical man was acculturated, sensitive to context and environment, attuned to real-world problems—he would have found common ground on this point with William Allan or Nash Herndon.[44]

The plan was to endow a chair for Lederberg, and to hire as faculty the super-bright graduate student Newton Morton, just finishing his Ph.D. under the population geneticist James Franklin Crow. Morton was good enough that neither Crow nor Lederberg wanted to lose him, and he was getting offers from other schools. The plan was approved in early 1957 and formalized in May. But nothing actually changed. No laboratory space was available at the medical school, although a new building was planned. So Lederberg and Morton remained in their offices in the Ag school's Department of Genetics, a medical Monaco surrounded by agricultural France. Lederberg left for Stanford soon after establishing the department (he won a Nobel Prize a few months later). Crow, also a Ph.D., assumed the chair; it was under him that the department actually moved into the medical school, hired more faculty, and grew into a viable program. Crow obtained a large grant from the Rockefeller Foundation and hired Robert DeMars in somatic cell genetics and virology and, in 1960, the biochemist Oliver Smithies and the cytogeneticist Klaus Patau. But the department's character remained true to Lederberg's vision of a basic-science program in a medical context. This was no heredity clinic.[45]

* * *

McKusick's division of medical genetics, then, was one of many institutes, departments, and divisions of medical genetics springing up around North America in the late fifties and sixties. Yet it was McKusick's that became the hub. His laboratory produced more than its share of important results, but no Nobel-caliber discoveries. Rather, the significance of McKusick's group has more to do with low-status activities, such as compiling, teaching, and promoting, than with glamorous scientific or clinical breakthroughs. McKusick was tireless in promoting the field, and he had a knack for parleying commonplace laboratory activities into institution-building enterprises.

For example, in the late fifties, he started a journal club for the members of the Moore Clinic to stay abreast of the field. Most lab heads have journal clubs. But McKusick took a comprehensive approach. Exploiting the size of his expanding group, he sought to collect, document, and summarize *everything* published relating to medical genetics in a year. In 1960, in Moore's old journal, the *Journal of Chronic Disease*, he published a two hundred–page review article called simply "Medical Genetics, 1959," with himself as the sole author and a long list of acknowledgments. (Today, all those acknowledged contributors would be listed as authors; multiauthor papers were less common then, and even McKusick's partisans admit that he tended to be autocratic about publication.) He followed that with three more successive annual reviews. He then compiled these data into catalogue form, which he first began to circulate as mimeographed copies. This was the first edition of *Mendelian Inheritance in Man*, often known as *MiM*, or "McKusick's catalogue." *MiM* was more extensive than Pearson's *Treasury of Human Inheritance*, and was encyclopedic rather than narrative in organization, but it served the same purpose. By 1965 McKusick had begun keeping *MiM* in computerized format for easy updating. Entries were built historically, like entries in the *Oxford English Dictionary*, and McKusick and his team updated them as new findings came in. In 1987 *MiM* went online, becoming *OMIM*, and in 1995 management and storage shifted from Johns Hopkins to the National Center for Biotechnology Information.[46]

Another of his discipline-building activities was the "short course" in medical genetics, taught at the Jackson Laboratory, in Bar Harbor, on the rocky coast of Maine's Mount Desert (accent on the second syllable) Island. The Roscoe B. Jackson Memorial Laboratory—later shortened to the Jackson Laboratory, or simply "Jax" or "Jaxlab"—was founded in 1929 as a cancer research facility. Clarence Cook Little, another student of William E. Castle's and a longtime eugenics advocate, was its first director. Under Little's stewardship, Jax had tackled the cancer problem through mice as a model organism, and it had built an international reputation in mouse genetics.[47] McKusick, a Maine native, was drawn to it as a place to do good science amid familiar beauty. In the summer of 1959, stopping through on the way to his summer house up the coast, he met with the Jackson Lab scientist Earl Green and John Fuller, the director. Returning to Baltimore, he sent round a memo to the Hopkins medical genetics group, describing the plan the three had hatched for a summer course, to be held the following year. McKusick was impressed that at the Jax, "they do in mice the same things that we do in human beings at the Moore Clinic . . . namely identify deviant phenotypes and figure out whether they are genetically determined and, if so, how they are inherited. Try to determine what the basic defect is and what can be done to modify the condition." The difference, of course, is that in mice you could do breeding experiments.[48]

The purpose of the course was frankly evangelical. "One of the leading functions of the course," he wrote in a letter to Fuller, "would appear to be in the recruiting line." It would be, in part, an effort to draw boundaries around their new field—an exercise in colonization. McKusick was partisan, a clinician to the end. "Some criticize the designation 'medical genetics' and favor 'human genetics,'" he wrote. "I do not agree, medical genetics states explicitly what we have in mind." At first, he envisioned genetic knowledge and enthusiasm trickling down from deans, advisers, and full professors to bathe the headcount junior faculty and students. "What I have in mind," he wrote, "is a unique course, aimed at medical school faculty—persons of instructor or assistant professor grade or higher, who want special instruction in genetics as an aid in their research and teaching." But after the first season, he had a two-tiered strategy: "Brain washing for recruiting: 1) Top level; 2) youngsters."[49]

McKusick used his network connections to raise funds to support the course. He served on the advisory board of the National Foundation for Infantile Paralysis (March of Dimes). The vice president for research at the National Foundation was the distinguished Rockefeller physician Thomas Rivers. "He was a Hopkins graduate," McKusick said. "I had sent him a copy of my book *Heritable Disorders of Connective Tissue*. . . . I think he was instrumental in getting me appointed to the Medical Advisory Board." Once on the board, in 1959 he spoke to Rivers and to Basil O'Connor, the founder and president, and arranged a grant. "The March of Dimes was the sole support of the course for its first twenty-five years," McKusick said, but that was not quite true. He also secured supplementary funding from the American Eugenics Society to supply student fellowships to cover tuition, travel, and lodging. The American Eugenics Society was only too happy to comply—so happy, in fact, that Frederick Osborn exaggerated its role, believing it had directly subsidized the course. Indeed, the Eugenics Society had a Medical Genetics Committee, which gave grants for coursework and research and sponsored conferences attended by respected, mainstream human and medical geneticists such as McKusick, Bentley Glass, and Lee Dice. In the mid-1960s it surveyed thirty-one medical schools and found that twenty-five were interested in learning how to incorporate medical genetics into their curricula. The Eugenics Society invited McKusick to give a talk on the integration of genetics into third- and fourth-year clinical teaching. Through the sixties, the Eugenics Society continued to have a presence in "serious" medical genetics.[50]

The Bar Harbor course seemed to slake a great thirst. McKusick received some two hundred applications for forty-five slots the first summer, and by 1965 it had doubled to ninety students. Excessive class size was a common complaint on otherwise strongly positive student evaluations. Lectures were held mornings and evenings, with afternoons free for special study opportunities, such as laboratories or tours of the mouse facility, or for leisure. The Bar Harbor course built on a long tradition of summer courses at beautiful seaside laboratories, such as Woods Hole, Massachusetts, Cold Spring Harbor, New York, and the Naples Zoological Station, in Italy. Wise organizers knew that the facility itself was a major draw. Families were typically welcome, and ample leisure time was built into the curriculum.[51]

The course quickly became a hub of McKusick's networking activities. He distributed copies of his massive reviews of medical genetics in the *Journal of Chronic Diseases*, and, in the mid-1960s, he gave each student a mimeographed copy of his new catalogue, *Mendelian Inheritance in Man*. He routinely nominated students in the short course for membership in the American Society of Human Genetics. He happily godfathered spinoffs at other institutions. "There is room for more of this sort of thing," he wrote to

6.3 Victor A. McKusick, networking at the Bar Harbor medical genetics summer course. Courtesy of Alan Mason Chesney Archives, Johns Hopkins University

the course graduate David Bonner in 1962, "and the idea for a course at La Jolla [California] seems like an excellent one. . . . I will be happy to mention this to this National Foundation people if you would like me to do so." The course also fostered collaborations between McKusick's group and those of former "students." "Please give my best regards to the other members of your staff," wrote the Harvard physician Park S. Gerald in a follow-up note after taking the course. "Have Ned [Boyer] drop me a post card when he can, telling of the results with the gorilla sera."[52]

McKusick had a sharp sense of marketing. After the second year, he pitched an article on the course to Tommy Turner, the dean of the School of Medicine: "I was thinking that a good story for the Johns Hopkins Magazine would be one on the Bar Harbor Course," he suggested, even noting the illustration possibilities: "The setting overlooking Frenchman's Bay at Oakes Center is, of course, very colorful." The magazine did end up sending a photographer to cover the course. A few years later, the March of Dimes suggested inviting some journalists to visit the course. McKusick went along with the idea, and in 1967 several science writers attended the last few days of the course. Special sessions were held in which researchers could present current findings in lay terms, and the press was encouraged to write them up.[53]

In 1968 the strategy backfired, when the outspoken population geneticist Richard Lewontin was invited to give the summation. "The fact of the matter is, I don't really know anything about human genetics," he began, disarmingly. He then launched into an outsider's critique of the field. In typical form, his intent was constructive, his style abrasive. He began with the recent suggestions, stimulated by a 1965 study by Patricia Jacobs as well as a couple of recent criminal cases from the news, that men with an extra Y chromosome have a tendency toward violence, aggression, and crime. An antisocial behavior chromosome. A full-blown national controversy over XYY males would occur in the mid-1970s, centered on Lewontin's colleagues at Harvard and MIT. But for now, Lewontin accepted the possibility of a crime chromosome for the sake of argument. "I mused on it and asked myself, from a clinician's standpoint or a social engineer's standpoint, 'Where am I after I have found out that some unfortunate five-year-old has an extra Y chromosome?' " The first question one needs to ask, he said, was whether one should tell the child or not. "Do I do him any good if I tell him?" he asked. "Do I increase the probability of his anti-social acts by telling him?" No one knew the answer, he said, but he made it clear that he suspected that the moral course of action for this and many other genetic conditions was "to keep your trap shut." Medical genetics still had too little to offer therapeutically, and there was the potential for considerable psychological harm. The organ theory of human genetics was simply too limited. Lewontin speculated on a future era of molecular medical engineering. "When the day comes, if it ever comes, when human disorders will be cured by fooling around with the messenger RNA, then indeed you'll want to have lots of genetical information, but so long as diabetes is treated by grinding up animals or producing insulin in some other way, the fact—which is in doubt, of course—that diabetes in one of its forms or other is inherited is not very interesting." In short, from a therapeutic perspective, medical genetics had not yet achieved relevance.[54]

Lewontin went on to describe lines of current research he thought *were* interesting, such as gene interactions in development, and emphasized that much about the genetics of behavior was extremely important and interesting. The distinction was lost on Judy Randal, a journalist in the audience. Randal apparently heard Lewontin's talk as an attack rather than as a critique. On August 15, her paper, the *Newark Evening News*, ran an article by Randal damning medical genetics as a field. "Last week at a genetics course on Mount Desert Island, Maine, a professor of zoology from the University of Chicago took issue with fellow scientists," she wrote. "About 100 doctors and others had spent two weeks there among the idyllic pointed firs learning about the arcane intricacies of genetics. Dr. Richard C. Lewontin told them that most of the information they were getting was 'not very interesting' and mostly 'a waste of time.'" McKusick and Fuller were furious, the March of Dimes leadership was chagrined, and for a while doubt was cast on the wisdom of inviting journalists to the short course. The practice continued, however, and to this day one can expect a flurry of news stories about human genetics every August.[55]

* * *

McKusick, then, is not considered the father of medical genetics because he made glamorous breakthrough discoveries. Certainly, he and his group were prolific producers of knowledge—McKusick's curriculum vitae boasts more than seven hundred publications. But the bulk of his contributions were case histories, disciplinary reviews, and pedagogical surveys—descriptive and synthetic, rather than analytical. He did eventually get an eponym—McKusick-Kaufman (also called Bardet-Biedl) syndrome, which occurs in about 2 percent of the Amish—and he made or supervised numerous contributions to the genetics of specific diseases. But a syndrome doesn't make one the father of a field. His epithet derives from his activities in the relatively low-status areas of compiling and cataloguing, teaching and mentoring, administering and organizing. McKusick made himself essential as a gatekeeper of knowledge, an impresario of intellectual exchange, a broker of personnel, reagents, and data.

Similarly, the expansion of medical genetics as a whole in the 1960s was founded on a foundation of humble technical advances, upon which were built prolific sites of knowledge production and professionalization. The greatest advances were in cytogenetics—a subspecialty launched by the combination of simple, preexisting techniques and the correction of a decades-old error. But cytogenetics provided mainly diagnostic benefits, and clinically speaking they were modest: "getting their organ" meant that medical geneticists could predict more accurately who would develop one of a small number of rare diseases. Admittedly, biochemical genetics was more complicated. For a few rare conditions, such as phenylketonuria (see chapter 7), an environmental trigger could be removed, say, through the diet. Nevertheless, in the overwhelming majority of cases, treatment in medical genetics still meant prevention—either curtailing procreation or inducing abortion after the fact.

What cytogenetics—and, to a lesser extent, the other subspecialties of medical genetics—provided was not therapeutic breakthrough but a sense of identity. They made it possible to think of oneself as a medical geneticist. And as soon as one could, many did. After the initial rush of discovery of "disease chromosomes" in the late fifties and early sixties, the most dramatic transformations of the sixties were pedagogical, administrative, and archival. Indeed, the activities that made McKusick the father of medical genetics constitute perhaps the most important aspect of 1960s medical genetics. The institutes, the courses, the journals and textbooks that mushroomed in the sixties were exactly the sort of professional infrastructure that Lee Dice, Laurence Snyder, Madge Macklin, and the medical geneticists of the thirties and forties had been striving for. The disciplinary infrastructure that McKusick, Motulsky, Crow, and others helped create made possible the advances that would soon place heredity at the very core of biomedicine. In the sixties, then, medical genetics came of age.

The fruition of the sixties emboldened this new generation of medical geneticists to new visions, new ambitions. Late in 1968, Roger Donohue, a postdoc in McKusick's lab, mapped a human gene—for the minor blood group known as "Duffy"—to a nonsex chromosome (chromosome 1). The same year, Torbjorn Caspersson at the Karolinska Institute in Stockholm finally identified banding patterns in human chromosomes using quinacrine mustard, which fluoresces when it interacts with DNA. Human gene mapping could at last begin in earnest. At the end of the sixties, human cytogenetics had finally caught up to where fruit fly genetics had been in the teens.

It took McKusick only a few months to leapfrog from one gene to a grand vision of the future. In 1969 he and the dysmorphologist Clarke Fraser organized a conference on birth defects, sponsored by the March of Dimes. McKusick gave the opening remarks. "Twenty-five years ago," he began, "a technical development beyond anything previously achieved by mankind—the harnessing of the atom—had been accomplished through the application of resources far in excess of those ever before applied to a single research project." That summer, he continued, man had landed on the moon, an "even more spectacular achievement," again attained with the application of huge resources to a rigidly directed research program. McKusick then called for a medical-genetic moonshot:

> I propose that detailed exploration of the genetic constitution of man is ripe for an all-out attack. The principles and broad outlines have been discovered. What we should know in full detail are the structure and geography of the chromosomes of man: the full nucleotide sequence of all genes determining the amino acid sequence of proteins—the so-called "structural genes"—and the location of each on the 24 chromosomes of man—the 22 autosomes and two sex chromosomes.[56]

A human genome project. It would be twenty years before it became a reality, and the sequence was achieved by methods and strategies inconceivable to researchers in 1969. Indeed, getting there ultimately involved the rejection of McKusick's clinically centered approach, his organ theory of genes, in favor of a radically new strategy rooted in biochemical genetics. That strategy at last began to fulfill the old dream of engineering the human body.

7
Genetics without Sex

AT THE CLOSE OF *THE DOUBLE HELIX*, James Watson's farcical memoir, Jim is wistful. Never mind that he and the dashing, flirtatious Francis Crick have just found what Crick breathlessly called the secret of life. For all that, he's lonely. DNA did not make him an immediate celebrity. Worse, the tweedy world of British science has proven bracing yet chaste. The only significant female character in the book is Rosalind Franklin, portrayed, untrue to life, as a priggish bluestocking. And though Crick is ever surrounded by beautiful women, Watson is repeatedly and comically snubbed. As the narrative closes, one suspects he is still a virgin. A celebratory posthelical holiday brings neither relief nor gratification. It is his birthday, he is in Paris, and he slogs the wet streets alone. The long-haired girls near Saint-Germain-des-Prés, wrote the hangdog Wunderkind, were not for him. "I was twenty-five and too old to be unusual."[1]

Sex was always the sticking point with genetics. Since Galton, scientists interested in human heredity had been dogged by the fact that one could not ethically carry out breeding experiments. The history of human genetics was one long quest for ways to circumvent the sex problem: extrapolations from animal models, mathematical estimates of the frequency of a given genotype in a population, pedigree analyses, and seizing upon "natural experiments," such as twins raised apart or exposure to atomic fallout. All of these methods yielded data, but all suffered from imprecision or a severely restricted dataset. Similarly, those who sought social and medical applications of heredity had only crude methods at their disposal: controlling marriage, controlling reproduction, controlling birth. Sex, in other words, connects the impulse of the individual to the needs of the population. A few visionaries had long imagined fanciful scenarios in which human genes could be manipulated experimentally, and human genetics could be regulated to eliminate or prevent disease, counter degeneration, or even make genetic enhancements. But as late as the 1960s, no practical means for such engineering were in sight. The scientific problem and the social problem were the same. Both required doing genetics without sex.

Like the girls on the Rue Saint-Germain, human-geneticists mostly ignored the twenty-five-year-old Watson—but likewise the rakish Crick. Nor were Watson and Crick especially concerned with the science or the practitioners of human genetics. The crystallography and physical chemistry undergirding the DNA structure were utterly alien to most researchers interested in human genetics or medical heredity. The professionalization of human genetics at midcentury was built mainly on low-tech cytogenetics; among the biggest breakthroughs of the 1950s were the use of deionized water and a tricky gesture with the thumb to spread chromosomes just so under the microscope slide. But while medical geneticists were finding their organ, developments in biochemistry and cell biology were leading to new workarounds for the problems of sex in modern genetics. In the fifties and sixties, the sex problem showed fissures, and then in the seventies and eighties it cracked, crumbled, and vanished with the invention of such techniques as somatic cell genetics, genetic engineering, and DNA sequencing. As the century closed, human cells and genes could be combined with each other—and with those of animals—at will. Sex was no longer an obstacle.

As that happened, Watson's career, which began with viruses, increasingly intersected with human genetics. Eventually he became an emblem of the medicalization of genetics. His racy memoir was only one element of a five-decade campaign on behalf of DNA. Through scientific, pedagogical, and popular writing, through administration and fund-raising, through networking, and through a lot of hard thinking, Watson devoted his entire professional life to installing the double helix at the core of biomedicine. His career featured technological and professional highs as well as embarrassing moral lows, much like the history of human genetics itself.

Eventually, he fell prey to the same sorts of prejudices and oversimplifications that have ever plagued this field. For better and worse, he is an emblem of the technical advances and the ethical pitfalls of the molecularization of medical genetics. Much happened that Watson had nothing to do with, but his career serves as a useful transect through the maturation of human cytogenetics and biochemical genetics, a process that marked the strong renewal of Garrodian approaches and established heredity at the core of the health sciences.

* * *

Genetics without sex required a kind of experimental synecdoche. In rhetoric, synecdoche makes the part stand for the whole. As one might refer to a monarch as "the crown," or as hospital staff might shorthand a cancer patient as "the lymphoma," so did cells and molecules come to stand for the entire person. By midcentury, researchers developed lineages of human cells on which they could perform a widening range of experimental manipulations that would be ethically unthinkable on whole human beings.

Culturing human cells proved much more difficult than culturing animal cells. Although Ross Harrison cultured frog nerve cells in 1907, it was not until the forties that a researcher—John Enders, at Harvard—could get human tissue to survive long enough in his laboratory to be experimentally useful. Then in 1950 Henrietta Lacks, a young African-American woman from outside Baltimore, entered the Johns Hopkins Hospital for treatment of cervical cancer. Her physician, George Gey, had been trying for years to culture human cells. He often biopsied his patients and tried to culture their cells. Today, such a procedure requires informed consent from the patient, but not so in the fifties; and, sadly, doctors then often gave even less thought to the rights of low-income patients such as Lacks than to their other patients. Gey sampled Lacks's tumor with neither qualm on his part nor consent on hers. Lacks's cells, however—lightly veiled in the literature as "HeLa" cells—were more than cooperative. They grew in dishes, on glass sheets, and in suspension. They grew over a wide range of temperatures. They traveled well. Gey sent them to all who asked. In Colorado, Theodore Puck learned how to isolate a single cell and grow up a vibrant culture from it—a so-called clone. That meant that slight variations—individual variations from cell to cell—could be isolated and studied. Soon there were many lines of HeLa cells, each with a unique character but all sharing a strong family resemblance. Like all model organisms, HeLa cells adapted well to life in the lab. They were low maintenance and reproduced (asexually) readily, and like all model systems they had experimental properties that made them useful for experimentation. Fruit flies threw many mutations; rats can be taught to perform complex tasks; dogs were particularly useful for physiology research. The advantage of HeLa cells was that they were similar to bacteria in practical and ethical terms, but they had human physiology.[2]

One thing cells could not do, of course, was have sex. Or could they? In 1946 Joshua Lederberg documented what he cleverly called sex in free-living single cells—bacteria. The sexual analogy is a loose one: bacteria have no male and female, there is no meiosis (the term, borrowed from rhetoric, to describe the halving of the genetic material at the formation of sperm and eggs), and the "sexual" act does not result in reproduction. But for Lederberg, these were details. The crucial point was the exchange of hereditary material, the basic act that makes genetic analysis possible. In 1952 he and his student Norton Zinder discovered "transduction," the transfer of genetic material from one bacterium to another by means of a virus. If bacteria could approximately have sex, so could HeLa cells.[3]

But the ever-prescient J. B. S. Haldane described the biology more precisely. After attending a meeting on bacterial and viral genetics where he heard about Lederberg's work among other research, he published what was, for him, a rather breathless review article entitled "Some Alternatives to Sex." In it, he noted that techniques were now available to achieve recombination in the absence of a true sexual process. "This observation," he wrote, "may be the key to human genetics." If one could arrange for a man with known genotype to have five hundred children, he continued, one could determine the linkage between two genes with good accuracy. "We cannot do this," he observed, "but we might be able to study his bone marrow cells in tissue culture" and analyze them. "By such techniques it may be possible to map the human chromosomes."[4]

In other words, technologies that accomplished alternatives to sex obviated the need for harsh, objectifying measures to control people's reproduction. The experimental synecdoche made possible everything from mapping chromosomes to treating disease to human improvement. It lowered the ethical barrier to human genetic manipulation for any purpose; and the various purposes to which people have always put genetic manipulation have always blended into one another. Genetics without sex enabled the pursuit of human perfection without the coercion of individuals that had made eugenics distasteful.

* * *

The double helix did not take the world by storm. The crystallography and phage communities, of course, read the DNA papers with interest. Watson became a star among his group, but outwardly his life changed little at first. He continued on temporary postdoctoral fellowships for the next three years, at Caltech and, briefly, back at the Cavendish, during which he tried the same approach with the related molecule RNA, with much less success. In 1956 Harvard University offered him an assistant professorship. The following year Mathew Meselson and Franklin Stahl confirmed the method of replication—called "semiconservative"—Watson and Crick had postulated, with what some called the most beautiful experiment in biology. Watson began to collect promotions and awards: associate professor (1958); the Lasker and Eli Lilly awards (1960); and, at age thirty-three, full professor (1961). In 1960 Marshall Nirenberg and Conrad Matthaei published the first crack in the genetic code, and the next year François Jacob and Jacques Monod published their "operon" model, which described the gene as an almost mechanical system. These developments gave the double helix real heuristic power. In 1962 Watson was elected to the National Academy of Sciences, and he, Crick, and Wilkins shared the Nobel Prize in Physiology or Medicine. The question of whether Rosalind Franklin—whose beautiful X-ray pictures were critical to solving the structures—would share in the prize was made moot by her death from cancer in 1958. Only after the Nobel did Watson and DNA become nationally known celebrities.[5]

Crick tended to avoid the limelight, but Watson enjoyed glamor and showmanship. Increasingly, he acted as DNA's scientific manager and literary agent. Once dismissive of biochemistry, he became proudly chauvinistic about the molecular approach as the one truly rigorous way to understand life. His Harvard laboratory attracted ambitious, competitive, celebrity-hungry students and fellows. Watson stayed out of the laboratory himself, preferring the role of impresario, mentor, and gadfly. The publication record underestimates his influence on the field, because he refused to add his own name to papers published by his students. This reflected research ethics, not humility. He turned awkwardness into bravado; refusing to cultivate a conventionally polished lecture style, he instead attracted crowds of students to disheveled lectures in which he mumbled thrilling field reports from the front lines of molecular biology, spiced with racy gossip about his friends and heroes. He relaxed his inhibitions and cultivated his idiosyncrasies. He viewed niceties of manners and tradition as obstacles to progress. He infuriated his colleagues, goaded his students, and chagrined the university administrators. To him, good science was ambitious, ruthless, and passionate.[6]

Watson became DNA's prime evangelist. A manifesto in textbook form, his *Molecular Biology of the Gene* appeared in 1965. Its authoritative yet informal style and bold, declarative subheads brought generations of students into the fold and set the standard for laboratory textbooks ever since. And then in 1968 *The Double Helix* appeared, having first been rejected by Watson's own university's press and then picked up by the new press Athenaeum. The book is structurally brilliant, a scientific coming-of-age story with all the pacing, humor, and plot twists of classical farce. Watson is pitch-perfect as the unreliable narrator, undermining our confidence in his accuracy. It enraged Watson's colleagues, particularly Crick, who felt it lessened the nobility of science, making it seem sleazy and shallow—a painful irony, since the book is a shoulder-chucking love letter to Watson's hero. As well, feminist scholars have been outraged by Watson's caricature of Rosalind Franklin, damning it correctly but simplistically as sexist. Watson, however, remained collegial with Franklin until her death, and she and Crick were close friends. Critically, then, the book's reception has been mixed. Yet it has found a huge and enduring audience. Young scientists, in particular, have loved the way it jettisons the stuffy ideal of science as an austerely noble pursuit. It replaces it with another ideal—equally contrived—of science as a playground full of fun and racy competition, where lazy, bumbling amateurs with big ideas can make a splash. It resembles no book so much as Paul de Kruif's sensational *Microbe Hunters* (1926), which it displaced as the popular work that lured young people into biomedical culture. The book became a best-seller, eventually selling more than a million copies.[7]

In his writing and public talks, Watson cultivated an aesthetics of DNA. Absorbing, doubtless from his mentor Max Delbrück, the physicist's sense of "elegance," or austere, rational beauty, he often wrote of "pretty" structures or ideas. In order to be scientifically pretty, something must be heuristically powerful and seemingly effortless. To be pretty is to be incisive, to solve a tough problem with few steps. Scientific beauty requires insight. The beauty of DNA stemmed from the combination of its curvaceous form and its extraordinary explanatory power. DNA was pretty from the outset, but the more one learned about it, the more beautiful, in this technical sense, it became. It appealed to a certain romanticism, an almost sentimental willingness, which many of Watson's group shared, to believe that simplicity reflected truth. Crick tempered his romanticism with a sardonic, self-deprecating wit, for example by referring to Watson's dictum that hereditary information moves from DNA to RNA to protein, but not back again, as the "central dogma" of molecular biology. Gunther Stent, a minor member and the self-appointed chronicler of the phage group, referred to the period 1953–68 as the Golden Age of molecular biology. DNA was alluring, its curvy lines replacing Eden's apple as the iconic depiction of self-knowledge. It permitted a new kind of genetics without sex, but the molecule itself was very sexy.[8]

* * *

While Watson scuttled around Kings College London trying to read Rosalind Franklin's X-ray images of DNA, nearby at University College London, Lionel Penrose was quietly reorienting human genetics. Born in 1898, Penrose grew up Quaker, like Francis Galton and Karl Pearson. He came from a stern, accomplished family, and had a strongly logical turn of mind. He read maths at St. Johns College, Cambridge, studying mathematical logic under Bertrand Russell. Taking a bachelor's in 1921, he traveled to Vienna to study psychiatry and met Sigmund Freud. In 1925 he returned to Cambridge to study medicine, taking his M.D. degree in 1930. The next year, he took a position at the

Eastern Counties Institution, in Colchester, Essex. Over the next seven years, he studied mental deficiency in the inmates of Colchester, using the standard eugenic and human-genetic tools of interviews and questionnaires, medical examinations, family histories, and pedigrees.[9]

Where H. H. Goddard had found a single gene for feeblemindedness, Penrose described a smooth continuum, from normal to barely functioning "idiot." He found a definite hereditary basis, but also significant environmental influences, including disease, maternal age, trauma, and infection. The genetics was complicated. With a warning to those who would use a simplistic germ theory of genes to justify policy decisions, Penrose concluded, "It has never seemed at all probable that a single cause could account for all mental deficiency in the same way that the *Spirochaeta pallida* accounts for all syphilis. The aetiology of mental defect is multiple and a facile classification of patients . . . would have only led to a fictitious simplification of the real problems inherent in the data."[10]

Penrose sat out the war in Canada, as director of psychiatric research for the province of Ontario. Near the war's end, Lee Dice nearly enticed him to the Heredity Clinic in Ann Arbor; the deal signed, Penrose backed out at the last minute to accept an offer to become the next Galton Chair of Eugenics at University College London. On assuming the chair, Penrose delivered the distinguished Galton Lecture. His subject was phenylketonuria (PKU). Described by the Norwegian physician Asbjørn Følling in 1934, PKU was by this time considered an inborn error of metabolism, related to the original inborn errors, alkaptonuria and cystinuria (see chapter 1). Penrose titled his lecture "Phenylketonuria: A Problem in Eugenics."[11]

Considered by many biochemical geneticists a manifesto of the replacement of eugenics by modern medical genetics—Barton Childs told me it ought to have been called, "Phenylketonuria: A Problem *for* Eugenics"—Penrose's Galton lecture is actually a bit more complicated. As he had in the Colchester Survey, in the Galton lecture Penrose rejected the germ theory of genes, writing, "We cannot take the same attitude here that we might with regard to some noxious pest and simply ask to have the offending genes exterminated." He looked back to the namesake of his chair, finding in Galton's writings a "broader view" of eugenics as the "steady though slow amelioration of the human breed." He rejected much of recent eugenics, however—especially its "propaganda aspect." He mocked the chauvinism and selectivity of the recent large-scale eugenic experiment on the Continent, noting that PKU, which seemed to be most prevalent in people of northern European ancestry, provided a "refreshing" contrast to Tay Sachs Disease (juvenile amaurotic idiocy), which is most common in Jews. "A sterilisation programme to control phenylketonuria confined to the so-called Aryans would hardly have appealed to the recently overthrown government of Germany," he observed drily. Yet he did not discount the importance of preventing mental disease. In the age of atomic weapons, mental defectives were an even greater threat than previously. "Now that weapons are constructed capable of instantaneous annihilation of large populations," he wrote, "the question of ensuring the intelligence and mental stability of people entrusted with power of decision has become extremely significant." Penrose thought mass sterilization a cruel and fruitless exercise, not least because one still could not yet identify carriers. He looked forward to this happening someday, and once it was possible, he thought marriage between carriers ought to be prevented. Nevertheless, PKU was among the best-known hereditary diseases of mental defect and a model for the kind of Galtonian eugenic practice he thought consistent with ethical medicine. Until other diseases were understood to the same degree, he wrote, "eugenic prognosis in the field of mental illness will remain, in most instances, a surmise based upon personal bias rather than a scientific judgment." The new Galton Professor, then, had a nuanced view of eugenics. He would not brook tolerance of crude and misguided coercive measures—they were both unjust and unlikely to have the expected consequences—but his interest in disease, heredity, and humanity could not allow him to lose hope for human-directed genetic improvement, however gradual.[12]

Although the first Galton Professor, Karl Pearson, had supervised the publication of the *Treasury of Human Inheritance*—an encyclopedia-style series of volumes on hereditary disease—the Galton Laboratory had long since lost any real relevance for physicians. Penrose refocused the Galton, not by turning it into a clinical facility but by taking up scientific problems important to physicians. Under Penrose and his colleague and successor Harry Harris, the Galton came to appeal to a new breed of physician-researcher interested in problems of human heredity.

"If you were going into medical genetics" in the 1950s, said the population geneticist James F. Crow, "part of the tour of duty was to spend some time with Penrose. As near as I can tell, the people that went over there didn't go there with any specific idea. They were just going to have tea with other people and with Penrose and pick up words of wisdom from the master." For Barton Childs, looking for a place to train in genetics in the early fifties, there was the Galton, there was the Michigan Heredity Clinic in distant second place—and there was no third. Neel had "just opened his place," Childs said, not remembering or not counting the fact that the Heredity Clinic had been run by Lee Dice since 1941. "It wasn't fully staffed." Nor was Childs fond of the prospect of living in Ann Arbor. "As opposed to London? I mean, where's the choice?" To him, at least in memory, the Galton had been "the one and only . . . fully developed department of human genetics" at the time. As researchers set up new institutes in the United States, they often toured the medical genetics and eugenics institutes of Europe to make contacts and find models. London was an obligatory stop on such tours, as were Copenhagen, Uppsala, Oslo, Paris, and for some, Germany. McGill University's Clarke Fraser made such a tour in 1954; Arno Motulsky took his three years later, when he was setting up the Division of Medical Genetics in Seattle; and Kurt Hirschhorn, from the Rockefeller University, made his in 1958. "During my tour of all these genetic centers of Europe," Hirschhorn recalled, "two people I *had* to visit in London were Lionel Penrose and Harry Harris." Hirschhorn ended up spending half a year at King's College London, where Harris was at the time. Harris and Penrose worked closely together, and Harris was the obvious choice to succeed Penrose as the Galton Chair when Penrose retired in 1965. The draw was the intellectual excitement, not posh laboratory facilities. "These were old buildings," recalled Charles Scriver. "When I went back to the lab at night and turned on the light, the floor moved. It was cockroaches."[13]

Penrose had a Garrodian perspective; for him, genetics was more powerful for describing one's constitution—diatheses, predispositions to disease—than for identifying specific disease agents. Harris was even more Garrodian. In 1963 he edited a reissue of Garrod's *Inborn Errors of Metabolism,* which he garbed in a loving biographical sketch of Garrod, an analysis of the *Inborn Errors,* and an essay on the " 'inborn errors' today" that advanced an intellectual genealogy of Garrod's vision and reviewed the abundant literature that could plausibly be framed as Garrodian. It is vintage scientific mythos: scientifically perceptive, heuristically powerful, overgenerous, and teleological. Harris's signal contribution was to link medical human genetics to the Watson-Crick structure and thus to provide a biochemical explanation for Garrod's inborn errors. "It is now generally accepted that the primary substance concerned in heredity is deoxyribonucleic acid (DNA)," he wrote, and that the functions of genes are specified "by the precise linear sequence of nucleotides." A mutation was now understood as a change or disruption in that sequence. An altered, damaged, or missing enzyme, for example, could disrupt a biochemical pathway, leading to a Garrodian inborn error. Thus one could now trace a direct and physical set of causal connections from genetic mutation to altered phenotype. Like Garrod—but unlike most followers of Garrod up until that point—Harris was interested in normal variation. Differences between individuals, he wrote, reflected differences in the enzymes they synthesize.[14]

But where Garrod's inborn errors were rare conditions, Harris examined common variants, or polymorphisms. Partly, this grew out of a limitation of his technique: in order for Harris to detect them, variations had to be fairly common—on the order of 1 percent or greater. He found striking polymorphisms in three of ten enzymes he examined. "Unless we have been excessively lucky in our choice of enzymes," he concluded, "polymorphism to a similar degree may be a fairly common phenomenon" in humans. What is the purpose of all this variation? he wondered. Do the frequencies of the different forms vary from one population to another? Are we glimpsing the material of evolution? What determines the selective pressure? Harris had begun to probe the subtle but ubiquitous biochemical variation that Garrod had found so compelling.[15]

Ironically, then, under Penrose and Harris the Galton became a hub for a new Garrodian genetics. As North American medical geneticists passed through the Galton Laboratory, they absorbed an approach to human heredity and disease grounded in biochemical variation, which they then brought back to the research hospitals of the United States and Canada. One effect of this emphasis on variation was to erode the difference between genetic disease and health. Variation was variation; some of it

7.1 Harry Harris, pioneer of polymorphism. Courtesy of Alan Mason Chesney Archives, Johns Hopkins University

was pathological, but much of it was not. As the rich store of human genetic variation came to be seen as something manipulable, the old distinction between correction of defects and genetic improvement began to disappear. The study of polymorphism opened up a new frontier in the science of human perfection.

* * *

Penrose and Harris understood what the graduate student Jim Watson initially had not: that biochemistry was the anatomy and physiology of single cells. In order for the cell to stand genetically for the human, it had to have traits. Beginning in the 1940s, biochemical techniques emerged that yielded rich descriptions of cells' anatomy and physiology, providing a store of cellular qualities on which biologists could hope to do genetics. One way to separate proteins was by their chemical affinity to a solvent. One would then visualize the proteins by staining them, which yielded a sheet with colored spots corresponding to the different proteins. Chromatography could be done on a tabletop with inexpensive reagents and filter paper. Using two different solvents at right angles enabled one to separate proteins in two dimensions.

The young Montreal physician Charles Scriver learned the technique of paper chromatography from Harry Harris. As a new M.D. at McGill University, still uncertain of his specialization, he stumbled onto a survey of the medical possibilities for paper chromatography by Charles Dent and John M. Walshe, colleagues of Harris's at University College Hospital. Like many of the advances in cytogenetics, paper chromatography was based on simple, even primitive principles. "We can only remain astounded," wrote Dent and Walshe, "that a modification [in technique] which now appears to us so obvious should have only come to us after such a long passage of time, and then only in a roundabout fashion." Scriver found chromatography fascinating, but foul. Solvent chromatography employed two noisome derivatives of coal tar: lutidine, judged by some to have the most horrid smell known to mankind; and phenol, a sickeningly sweet and tarry compound found in poisonous mushrooms and bad beer. Scriver said his wife always knew when he had been doing chromatography that day, from the reek of his clothes.[16]

Another approach was to separate proteins by electrical charge. In 1937 in Uppsala, Sweden, Arne Tiselius, a student of Theodor Svedberg, inventor of the ultracentrifuge, had invented the technique of electrophoresis, which enabled him to separate the protein components of biological fluids based on differences in electrical charge. The Tiselius apparatus was biology's answer to the cyclotron. It was twenty feet long and five high and cost thousands to build and thousands more per annum to operate, as well as a staff of skilled technicians. In the late forties, few universities could afford a Tiselius. The California Institute of Technology had one. They also had the great physical chemist Linus Pauling. Using the Tiselius apparatus, Pauling found an electrophoretic difference between the hemoglobin in normal patients and those who have sickle cell anemia. Like Lederberg, Pauling had a flair for language and showmanship: he called sickle cell a "molecular disease." The same year, Jim Neel at Michigan sorted out the genetics of sickle cell, deciding that it was a simple Mendelian recessive, with the milder, almost asymptomatic sickle cell trait apparently the heterozygous condition and sickle cell disease the homozygous recessive state. In 1959 Vernon Ingram identified a single amino acid difference between normal and sickle hemoglobin; sickling was caused by a single mutation. Human biochemistry and disease genetics were linked.[17]

Researchers played with new methods to try to improve the speed and lower the cost and effort of electrophoresis. In 1951 Tiselius and his student Henry G. Kunkel developed paper boundary electrophoresis, in which the protein solution is pulled across a sheet of filter paper and the bands are "frozen" in the paper, resulting in discrete spots that could then be cut out, purified, and analyzed. "You could put a lot of stuff on a big piece of paper, turn on the power and let it flow, and then stain it," said Charles Scriver. "And you could separate amino acids by distance in these runs." Solvent chromatography and electrophoresis each had advantages and drawbacks. Dent's group used both. "The nice thing about the electrophoretic approach," Scriver said, "was that it maybe caught on fire once in a while, but it didn't have a terrible odor."[18]

Filter paper electrophoresis was hampered by low capacity, low yields, and the tendency of some proteins to stick to the paper, which made separation difficult. Searching for alternative media to replace paper, researchers experimented with a wide variety of materials, including glass beads, glass powders, sands, gels, resins, and starch—both potato and corn. Henry Kunkel used powdered potato starch, available at any grocery. His method was much more compact and inexpensive than the Tiselius free-solution method. Although compact enough to fit on a tabletop, it could be used to analyze a fairly large sample.[19]

Still, electrophoresis with starch grains had its drawbacks. "It was quite laborious," said Oliver Smithies, who learned the technique in the early 1950s, at the Hospital for Sick Children in Toronto. "You would make a tray into which you could pour this slurry of starch and electrolyte and salt solution and let it settle. Then you'd get this damp mass of starch grains, . . . like a sand pie." You would then run electric current through the tray the long way and allow proteins to migrate along the gradient, stopping when they reached electrical equilibrium. "To find the protein," Smithies continued, "you couldn't see it, so you had to cut the damp starch slurry into the best you could do for slices, and do a protein determination on each slice." He enjoyed telling the story of thinking back to when he was a boy in Yorkshire, helping his mother do the laundry. "She would cook starch up, starch powder, and make a sort of semi-gelatin-like material, and dip my father's shirt collars into this and then iron them. Then when it was left around it would set into a jelly, I'd noticed. So I thought, well, if I take the starch and cook it, I can make it into a jelly, and then I can stain the jelly, because that will stain without falling apart."[20]

Cooking the starch made all the difference. Smithies's starch gel electrophoresis democratized protein analysis. It was less bulky and less expensive than any other method except paper electrophoresis—and it achieved much higher yields than paper. It could be carried out by a single researcher alone in the lab on Saturday night. As with the squash technique or deionized water for chromosomes, the power of the technology derived not from its sophistication but from its simplicity, its accessibility. Anyone could now analyze complex proteins. Other electrophoresis techniques were introduced, such as polyacrylamide gels and "molecular sieve" electrophoresis, but starch-gel electrophoresis continued to be among the most widely used techniques of biochemical genetics into the 1970s. Like the handful of little techniques that Tjio and Levan combined to correct the human chromosome number, cooking the starch was a minor, low-tech innovation that snowballed, transforming the ideas and culture of science.[21]

* * *

"Biology is undergoing a revolution whose meaning and magnitude have become apparent only in recent weeks," crowed the *New York Times*. It was February 1962. The old eugenic motto had been "the self-direction of human evolution"; the new genetics was making such control foreseeable. The center of the revolution was not the double helix but the genetic code. The code was the Rosetta Stone of biology: with it, one could interconvert

DNA, RNA, and protein. Biology was being reframed in terms of information. Genes stored information—assembly instructions and parts lists—about proteins, the often large, complex molecules that compose enzymes, receptors, transporters, and other vital components of cells. The code would specify how information stored in the genes would be "read" into protein. The mysterious step was going from the four-letter alphabet of RNA (adenine, guanine, cytosine, and uracil, abbreviated A, G, C, U) to the twenty different amino acids that are the building blocks of proteins. Researchers knew that small molecules called transfer RNAs read the RNA message in groups of three. Four nucleotides can combine into triplets sixty-four different ways. Each triplet must specify an amino acid. What was the code by which nucleic acid triplets were translated into protein sequence?[22]

Nirenberg and Matthaei constructed a monotonous artificial RNA message (UUUUUU . . .). They added it to Paul Zamecnik's "cell-free system"—a recipe of cellular components and juices for studying protein synthesis in a test tube (genetics without cells!). And they recovered an amino acid chain of pure phenylalanine. The RNA triplet UUU codes for phenylalanine. This early success prompted hopes that all sixty-four possible triplets would soon be matched with all twenty amino acids. Crick said it ought to be done in about a year. (It took five.) And then, in 1961, the French bacterial geneticists François Jacob and Jacques Monod described the machinery of the gene encoding the enzyme lactose (*lac*). It has, they showed, an architecture: the coding region, which contained the specification of the molecule's amino acids, was flanked by regulatory regions, the operator and the promoter, which do not themselves code for protein. The whole affair had a kind of mechanical structure, with what one could imagine as switches, dampers, and governors. Molecular biologists seemed to be down in life's engine room, studying the machinery and drawing up blueprints from it. The *Times*, in a forced metaphor with the Manhattan Project, gushed that "biological bombs" could soon drop. Besides determining the "basis of thought" and developing therapies for incurable diseases such as cancer, biologists might soon be "controlling the inheritance, and hence the destiny" of plants, animals, and humans, and even "creating life" out of bottled chemicals. The prospect of engineering the genetic material immediately, perhaps naturally, leads to the idea of controlling and perfecting the human animal.[23]

Biologists immediately began thinking about using these new technologies to control, cure, and even eliminate disease. Austin Weisberger of Western Reserve School of Medicine in Cleveland incubated sickle cell DNA with normal immature human red blood cells from bone marrow; when the cells matured, they sickled. In the converse experiment, cells from sickle cell patients were "rescued" and did not sickle. Also, Waclaw Szybalski, a Polish researcher who had worked at Cold Spring Harbor in the early 1950s and had seen Watson, in short pants, present the double helix to the summer crowd at the 1953 Cold Spring Harbor Symposium, heritably transformed a biochemical trait in HeLa cells. He and his wife Elizabeth incubated cells unable to produce an enzyme required for DNA synthesis with naked DNA extracted from cells that could produce the enzyme. The cells acquired the trait. And when they divided, their "daughter" cells retained the trait. Under the right conditions, simply mixing cells with naked genes could create heritable changes in the cells. Victor McKusick acknowledged the stirring medical possibilities opened up by the prospect of the manipulation of human genes. Discussing the research in his year-end review of medical genetics, he wrote, "The therapeutic possibilities in hereditary disease and in neoplasia stimulate the imagination." For the first time, researchers had a concrete mechanism with which to imagine "correcting" disease genes.[24]

The double helix became the emblem of the new molecular genetics. Struggling to explain the technicalities of the science, glorify the scientists, and sell newspapers, the media ascribed practically all vital activities to the Watson-Crick structure. DNA, wrote a *Times* reporter, was leading to a "revolution far greater in its potential significance than the atomic or hydrogen bomb." DNA "determines whether the fertilized human egg (ovum) will be born a Newton, a Beethoven, an Einstein, an 'average man' or even an idiot." Studying an individual's DNA, speculated the American Medical Association, might enable the discovery of chemical changes that lead to weakness or disease. Corrections might then be made, with drugs or by "juggling the code itself." It was understood immediately that a chemical description of disease tends to lead to chemical solutions, either through pharmacology or through what came to be called genetic engineering and gene therapy. The Nobel for the double helix later that year, then, was symbolic of this revolution in biology; it was an anointment of the molecule and the explanatory strategy that lay at the heart of these visions of human engineering.[25]

These fundamental findings raised a great many ethical questions for human genetics. As they had in the wake of the atomic bomb, scientists and other academics organized conferences with ominous-sounding titles—*Man and His Future* and *Biological Aspects of Social Problems.* In 1963 Joshua Lederberg noted that the new genetics had revivified classic eugenic goals. "Ultimately," he wrote, molecular medicine could "diagnose, then specify, the actual DNA composition of the ideal man." Yet emphasis on eugenics as the point of application of molecular biology overlooked the striking potential for what he called "developmental engineering," or euphenics. In time, he thought, euphenics would lead to an improved, idealized, and truly capable eugenics, but in the meantime, much useful direct application of biology could be made through understanding and regulating the development of individuals rather than attempting the much more challenging control of evolution. Lederberg clearly had sympathy for eugenic concerns, although he thought modern-day eugenicists were thinking too far in the future. In 1966 he labeled fantasies of direct manipulation of human genes "algeny," a portmanteau of "alchemy" and "genetics." Algeny was "diversionary," he wrote, not because he doubted it would come to pass eventually, "but because the obvious difficulties provide a too convenient refuge for evading sooner anxieties."[26]

Lederberg's terminology was never widely adopted, but his strategy was. Euphenics brought biological control of heredity down to a time scale that biologists and doctors could work with. Further, it was the missing link between the individual and the social. Euphenics is what enables medicine to address individual and family well-being in the immediate term without giving up larger aspirations to human perfection. Subsequent developments in molecular biology and genomics may be called biomedical advance by supporters or eugenics by detractors; however, much of it is better described as euphenics, the effort to shape both health and human nature by governing the processes of growth, development, and aging. The main strategies of genetic medicine since the 1960s have been euphenic, although algeny—we came to call it gene therapy—has always had its cheerleaders and its cowboys.

In light of such considerations, the American Society of Human Genetics formed an ethics committee to document, analyze, and discuss the many ethical questions raised by the new genetics, and they appointed James V. Neel to chair it. In a memo, he listed a wide range of topics to be addressed. Some of these were "conventional" topics independent of engineering approaches to biology and genetics without sex—Lederberg's "sooner anxieties." The physician Carl Djerassi had introduced the birth control pill in 1960, separating sex from reproduction—sex without genetics!—and thus opening up a basket of eels concerning reproductive choice and possible eugenic programs. Besides its cultural impact upon marriage, child rearing, and reproductive rights, the Pill was also euphenics: "therapeutic" birth control could have an immediate medical rationale and still quietly address larger questions of population control. And in 1962 the physician Robert Guthrie advanced Penrose's agenda in the study of PKU by introducing a simple blood test for the condition. The "Guthrie card" was inexpensive and, while not fool-proof, relatively easy to complete. It meant that even small-town general practitioners could test their patients for PKU. Screening programs were rapidly put in place across the United States—and, using a test developed by Charles Scriver, in Canada. Although PKU was generally understood as a form of mental retardation rather than as a genetic disease, the geneticists appreciated the social impact of newborn and prenatal screening for hereditary traits. Neel's list included genetic screening and counseling, adoption, artificial insemination, abortion, and fetal diagnosis, as well as the dysgenic effects of medical practices and the genetics of race as subjects of current concern in human genetics.[27]

More striking still, Neel went on to list subjects of future concern. These depended on genetics without sex and bioengineering. Heading the list was "germinal choice (Muller's ideas)." First put forward in Hermann Muller's 1935 *Out of the Night,* germinal choice was a scheme for eugenic improvement, involving highly selective sperm banks for genetically well-endowed donors, combined with artificial insemination or in vitro fertilization. Besides in vitro fertilization itself, Neel indicated the deliberate selection of particular gametes, such as sorting X and Y sperm to choose the sex of a baby, as well as parthenogenesis, vegetative reproduction (organismal cloning), the creation of "mosaic" individuals through cell hybridization, and "genetic surgery"—algeny—including modification of the DNA through transduction, DNA replacement, or the replacement of whole chromosomes. None of this was possible, but all of it was imaginable, in increasingly concrete terms. As the sixties drew to a close, human-geneticists were cultivating a vision of preventive medicine and human improvement by direct manipulation of the hereditary material.[28]

* * *

In his Harvard laboratory, Jim Watson was at the epicenter of the new genetics. In 1967, working with the phage called *lambda,* his student Mark Ptashne isolated the long-sought repressor, a protein that binds to the operator region on the DNA, inhibiting the reading of the gene. Two years later, Jonathan Beckwith, a young faculty member in bacteriology, along with students and fellows, notably James Shapiro, isolated the lactose operon in a test tube. In intellectual, left-wing Cambridge, the result was stunning scientifically but troubling ethically—to no one more than the researchers themselves. Once you could cut a gene out, it would not be long before you could paste it back in somewhere else. Would scientists soon be mixing and matching genes, blurring species boundaries or creating the master race? Beckwith held a press conference to warn the public about the implications of his own research. "The work we have done may have bad consequences over which we have no control," he said, drawing a parallel to the development of atomic energy. In this era of student activism against the Vietnam War, Beckwith's concern was Orwellian governmental abuse of this powerful new technology. As Shapiro put it, "The use by the Government is the thing that frightens us." Taking the cue, the *Times* called the research a "step in heredity control."[29]

Cambridge became a hotbed of science activism. Beckwith and Shapiro were soon joined by Richard Lewontin, recently arrived from the University of Chicago, Jon King at nearby MIT, the physicist Charlie Schwartz, and others in forming an activist group called Science for the People. Watson, too, joined in the spirit of bioactivism. Writing in the *Atlantic Monthly* in 1971, he opined circumspectly that a "blanket declaration of the worldwide illegality of human cloning" might be worth considering.[30] There is a species of genetic determinism bound up with much criticism of the new biology.

In spinning their attention-getting, worst-case scenarios, critics tend to overdramatize, to overestimate the power of their science, and to oversimplify the causal chain from basic biology to global domination. The work has hazards, to be sure. But they are often not those that one initially fears.[31]

Watson started to pull away from Cambridge. In 1968 he became director of his beloved Cold Spring Harbor Laboratory. Years of neglect, mismanagement, and valiant but insufficient or ill-advised effort had left "the Lab" a shambles. Broke, decrepit, and almost empty, it needed to be reinvented or shut down. Watson had served on the board of trustees since 1963. At a dire board meeting in 1967, he delivered a high-decibel discourse on the action someone needed to take in order to save the Lab. Bentley Glass, chairman of the board, recalled someone, possibly Norton Zinder, saying, "Jim, if you know just exactly what ought to be done, why don't you take it on?" Watson, Glass continued, gave a characteristic pause and said, in a musing voice, "Maybe I will."[32]

Watson remade Cold Spring Harbor in his own image. He rejuvenated the place, fashioning it over the subsequent years into the administrative control center, the intellectual expression, the physical and even aesthetic manifestation of his vision of DNA as the most important molecule in the world, and of research on it as the world's highest calling. The engine of the Lab's recovery was cancer research. Since the mid-1950s, Watson had been interested in so-called tumor viruses—any of several groups of viruses that cause either benign or cancerous tumors. Following the phage theory, even if the viruses themselves did not turn out to be major sources of environmental cancer, the compact, discrete genome of a virus could be a powerful way to strip the complexities of cancer down to their purest, most atomic form. In 1964 the National Institutes of Health initiated a Special Virus Cancer Program, on which Watson became a consultant. Among Watson's first acts as director was to hire a couple of tumor-virus geneticists to enliven the place.[33]

The Lab was thus well-positioned in 1971 when President Richard Nixon signed the National Cancer Act. The result of a compromise on a plan that would have created a cabinet-level federal cancer agency, the Cancer Act of 1971 instead elevated the National Cancer Institute—the first and largest among the various branches of the National Institutes of Health—to quasi-cabinet-level status. The NCI director was now a direct presidential appointee whose budget went directly before the president. Through the early seventies, while the budgets of the other institutes fell, NCI's rose dramatically, increasing from $180 million in 1970 to $762 million in 1976. A large public relations program, which Nixon called the War on Cancer to distract the public from the unpopular war in Vietnam, rallied public support for congressional appropriations. Watson tapped these new funds to rebuild Cold Spring Harbor. Besides large federal grants, he courted wealthy neighbors and supplemented the Lab budget with his own book royalties. He rehabilitated buildings, built new laboratories, enlarged the campus, and expanded the educational programs, such as the venerable annual symposium and the program of focused summer courses. One may debate the successes and failures of the War on Cancer for medicine at large, but it saved Cold Spring Harbor.[34]

In one of those summer courses, the 1971 edition of animal cell culture, Janet Mertz, a Ph.D. student under Paul Berg at Stanford, mentioned in a discussion on research ethics that her lab had all the pieces in place to perform a type of genetic engineering experiment. For years, Berg had studied another operon, known as *gal*, which codes for the sugar galactose. He was now taking the next step that Beckwith and Shapiro had feared: he had cut the *gal* operon out of the tumor virus SV40 and wanted to insert it into the bacterium *E. coli*. *E. coli* is the bacterial white lab rat; but it also occurs "in the wild": the human gut. Mertz had been an undergraduate at MIT in the late sixties, knew of Beckwith and the biologists' political activism brewing in Cambridge. She had a pang of leftist concern about government-controlled genetic engineering. She justified her decision by saying that the Berg experiment was "still a long way from being able to change people's genes." Going to Stanford, she said, "was sort of a compromise between my radical point of view and wanting to work on specific types of projects and have Paul Berg as my thesis advisor." Passionate about science and lukewarm about politics, Mertz compromised by doing the science she wanted to do.[35]

Bob Pollack had a less ideological concern: what if the bug escaped? He was one of Watson's new staff scientists and an instructor in the course. Pollack told Watson of his concerns, and then he telephoned Berg out at

Stanford. A meeting on biohazards—including but not limited to recombinant DNA—was organized at the stunning Pacific campground and conference center Asilomar, near Monterey, California. Another small conference in New Hampshire in 1973 led to a call for a national committee to explore the ethical and safety issues. In July 1974 the National Academy of Sciences Committee on Recombinant DNA Molecules Assembly of Life Sciences published its findings in the pages of *Science*. The committee had Berg as chair, and included Watson, Norton Zinder, and nine other leaders in the new research. "Scientific developments over the past two years," they wrote, made it "both reasonable and convenient" to make recombinant DNA molecules that blended genes from different organisms and different species. They continued,

> Although such experiments are likely to facilitate the solution of important theoretical and practical biological problems, they would also result in the creation of novel types of infectious DNA elements whose biological properties cannot be completely predicted in advance.
>
> There is serious concern that some of these artificial recombinant DNA molecules could prove biologically hazardous.[36]

The Berg Committee made the remarkable recommendation that the biological bomb of recombinant DNA should not be dropped without further research into safety. Self-consciously seeking to avoid the post-hoc hand-wringing of the atomic physicists in the fifties, the biologists recommended self-imposing a moratorium on recombinant DNA research until the potential hazards were better understood and better mechanisms were in place for regulating the research and preventing the spread of possibly dangerous organisms. In Cambridge, Science for the People worked the community into a lather. The town council passed an ordinance banning recombinant DNA research outright within the city limits. If recombinant DNA research was a horserace (as Watson saw it), this knocked two of the strongest stables—Harvard and MIT—out of the running. Top researchers fled for other institutions—including Cold Spring Harbor—where they could continue their work.[37]

A second Asilomar conference, in February 1975, brought together Watson, Berg, Lederberg, and 150 other researchers to discuss the moratorium and the hazards it was intended to alleviate. Memorialized as a gesture of ethically motivated experimental restraint, the consensus from Asilomar II was actually to *lift* the moratorium, replacing it with regulation. The meeting established physical and biological standards of containment (self-consciously borrowing the Cold War nuclear rhetoric) and called for the creation of a federal Recombinant DNA Advisory Committee (RAC) to oversee the research. When the Asilomar recommendations appeared, Science for the People immediately published an open letter to the conference participants, observing that the original criteria for ending the moratorium had not been met, and strenuously requesting that the moratorium remain in place. By then, however, the sea had changed: the researchers wanted to get back to work. Watson, who helped lead the discussion to impose the moratorium, became one of its most outspoken critics. When, at a Washington hearing, a member of a group critical of recombinant DNA asked him, "How can you let me down? You scientists have created this issue, and you should keep it going," Watson replied, "Because I was a jackass is no reason for you to continue to be one."[38]

Just as the recombinant DNA debate was dying down, Beckwith, King, and Science for the People learned of another troubling experiment happening on the Harvard campus. Its roots lay in the cytogenetic studies, beginning in the late fifties, that linked specific diseases to aberrant chromosome number. Among the descriptions of Klinefelter (XXY) and Turner (XO) syndromes, the Edinburgh group had discovered several individuals with an extra Y chromosome—so-called XYY males. In 1962 William Court Brown suggested that XYY males might have exaggerated male tendencies toward aggression and violence, and in 1965 his colleague Patricia Jacobs published an article suggesting that this was in fact the case. The public picked up on the story, which temptingly seemed to reduce a vexing social problem to a simple structure clearly visible under a microscope. Richard Speck, a mass murderer recently in the news, was reported to be an XYY male. Perhaps the Y chromosome was the "criminal chromosome." It was the germ theory of genes writ large. Dick Lewontin had mentioned XYY males in his 1968 Bar Harbor lecture as an example of the fruitlessness and ethical murkiness of much human genetics research (chapter 6).[39]

The controversial study teamed an experienced medical geneticist with a psychiatrist. Park S. Gerald was an M.D. hematologist who, like Arno Motulsky, had been attracted to biochemical genetics and in the early sixties had become a regular faculty member in McKusick's summer course at Bar Harbor. Gerald had teamed up with Stanley Walzer, a faculty psychiatrist, and had for years been pursuing a long-term study on patients with unusual patterns of sex chromosomes, such as XXY (Klinefelter syndrome), XO (Turner syndrome), and XYY. A major study, funded by the National Institutes of Health and published in 1969, was classic descriptive cytogenetics: Gerald and Walzer examined the chromosomes of twenty-four hundred phenotypically normal patients to ascertain the rates of sex chromosome anomalies. They found four cases of an abnormal number of sex chromosomes (aneuploidy) and nine more of chromosomal rearrangements. They then planned an expanded study to screen a larger group and analyze their chromosomes; Walzer would follow up with long-term observation and counseling. Walzer framed it as a test of hypotheses such as Jacobs's, which he and Gerald saw as both unquestioned and untested. Researchers, Walzer said in an interview, had "made assumptions that the XYY causes violent behavior and anti-social behavior. Which, of course, is horseshit. But that's the way the belief structure was for a long long time. So then a few of us around the country decided we were going to get a cohort of newborns and follow their development because there's so much bias otherwise."[40]

When Beckwith learned of the experiment through a colleague, he became furious. Although the study included all chromosomal abnormalities, he focused on the XYY males; they were the only ones linked to antisocial behavior. He had been reading the history of eugenics and he saw parallels between the idea of a crime chromosome and Henry Goddard's conclusion that feeblemindedness was a single-gene Mendelian trait. Both reflect a deterministic, germ-theory-of-genes approach to heredity and antisocial behavior. Beckwith and King saw the Gerald and Walzer experiment as a self-fulfilling prophecy. There was no ethical way to do such an experiment, they said. If you screen people and don't tell them their status, you're withholding medical information; if you do tell them, you're biasing your results. Through protests, demonstrations, speeches, and articles, Beckwith and King aroused opposition to the study, and eventually Harvard asked them to stop the screening portion of the study, although not the portion involving long-term care and monitoring. Science for the People has a mixed legacy. Although it often oversimplified the science and romanticized the virtues of making science policy by democratic consensus, the group was keenly aware that evaluating the merits and impact of the new genetics required more than biological training. Science had spilled over its banks into politics, sociology, ethics, and economics. The impact of the new genetics, members of Science for the People believed, was too great to be left entirely in the hands of the experimenters themselves.[41]

* * *

The new genetics was producing a new, universalist, molecular view of life. The genetic code, which translates the As, Cs, Gs, and Ts of nucleic acid into the amino acid chains of proteins, turned out to be strikingly consistent, almost universal, across the organic world. Up to the level of the cell, it became possible to combine, fuse, exchange, and hybridize parts from wildly different organisms. The part of sex that is interesting to a molecular biologist became increasingly easy to achieve in all imaginable combinations.

In 1960 Georges Barski, Serge Sorieul, and Francine Cornefert created the first hybrids of somatic (nongamete, "body") cells.[42] In the mid-sixties, hybrids of mouse cells with cells of rat and hamster were created, and in 1967 Mary Weiss and Howard Green, in the pathology department at NYU School of Medicine, hybridized mouse cells with human cells. The resulting culture "failed" productively: the way in which it didn't work jump-started human gene mapping.[43]

If you grow two different cell types together, about one in a million cells of one type will fuse with a cell of the other type. In Weiss and Green's experiment the human chromosomes were slowly eliminated as the cells continued to divide. Cell division by cell division, the hybrid reverts to a mouse cell. This property made the hybrids a unique system for mapping genes. "It's a gene transfer system," Frank Ruddle, one of the pioneers of the technique, told me. "You're putting all of the human chromosomes into this hybrid cell, and then you're losing them randomly—they leak out. But the fact that you can produce this hybrid is, in a sense, like in the bacterial system, taking DNA and adding it to cells, having them pick it up and be modified in some ways, according to the DNA that you've introduced." Weiss and Green realized that the "leaking" of human chromosomes allowed researchers to localize a trait to the chromosome that was lost: if you were following a trait of the cells, such as the ability to thrive without the addition of a particular nutrient, and that trait were lost at the same time a given chromosome was lost, the trait could be linked to that chromosome. It was Haldane's 1954 prediction come true, with a bizarre trans-species twist that even he did not foresee.[44]

Somatic cell genetics provided researchers with a kind of mock sex, which enabled them to carry out manipulations that would be taboo or physically impossible with whole organisms. Scientists could "mate" any two individuals they chose: one species to another, a human to an animal, or even one individual to itself. Like crossing-over at meiosis in conventional genetics, somatic cell genetics enabled genes to be mapped to chromosomes and even to sections of a chromosome, as well as ordered relative to one another along the chromosomes—just as Alfred Sturtevant had figured out with *Drosophila* in 1913. Moreover, Caspersson's introduction of chromosome banding at the end of the sixties made it possible to distinguish the various human chromosomes from one another and to impose a topography upon individual chromosomes. At last it had become feasible to create maps of human chromosomes.

Using these techniques, Ruddle's group at Yale began systematically mapping human genes. They soon began collaborating with Victor McKusick's group, from Johns Hopkins. Dirk Bootsma, from Erasmus University in Rotterdam, also joined in early on. The effort seemed to justify a workshop to coordinate efforts and chart progress. "I felt that since the mapping was going to proceed at a fairly rapid clip once the techniques were worked out," Ruddle said, "it would be good to have annual or every other year conferences to keep track of the emerging map." The first biennial International Gene Mapping Workshop was held in New Haven in 1973. The meeting then rotated to Baltimore and Rotterdam and back around again. The Gene Mapping Workshops tracked the nearly geometrical expansion of the maps. In 1973 researchers had mapped nearly 100 genes; at the 1977 meeting they had 210. By 1983 the number was over 450. By 1991, just nine years after the first Workshop, 2,325 genes had been mapped. Early on, Ruddle created a database to catalogue and track progress, and linked it to McKusick's catalogue of genetic diseases, *Mendelian Inheritance in Man*.[45]

The link between Ruddle's and McKusick's catalogues was more than collegial. Both were based on the unquestioned assumption that the only logical way to gather gene data was to start with a trait and then locate, and ultimately sequence, the gene. There is, then, a residue of the germ theory of genes in the mapping effort—an impulse to find the hereditary agents of disease, in contrast to seeking the hereditary components of constitutional response to disease. The mapping effort evolved into an international Human Genome Organization (HuGO), which in the 1980s made significant contributions to the international, multi-institutional Human Genome Project. HuGO's stance was that sequencing resources ought to be conserved by focusing on known genes and ignoring "junk" DNA such as noncoding sequences and long, monotonous repeats. It was a logical position, so long as sequencing cost and effort were the rate-limiting steps in the genome project. In 1987 McKusick, Ruddle, and Tom Roderick from the Jackson Laboratory in Bar Harbor were having drinks at the now-defunct McDonald's Raw Bar in Bethesda, Maryland, when Roderick suggested *genomics* as the name of the field of gene mapping and sequencing and the title of their proposed journal. Unlike the process of naming the *American Journal of Human Genetics*, there was no searching, philosophical debate. *Genomics* it was.[46]

* * *

Up to this point, "mapping a gene" actually meant localizing a *trait* to a spot on a chromosome. In the early days, the presence or absence of a physical characteristic was linked probabilistically to the behavior of a chromosome or chromosome segment. With bacterial genetics and somatic cell genetics, the trait became synonymous with a molecule, a gene product, but the points on the maps remained abstractions.

The nature of mapping changed fundamentally in 1980, when David Botstein, then at MIT, Ray White, from the University of Massachusetts medical school in Worcester, Mark Skolnick, from the University of Utah, and Ronald W. Davis, from Stanford, developed a means of using DNA sequence itself as a marker. Recall Harry Harris's study of enzyme polymorphisms; Botstein and colleagues developed an elegant way to find polymorphisms in the sequence itself, based on an observation two years earlier by Y. W. Kan and André Dozy, working once again with blood, and specifically with sickle cell anemia. They used restriction enzymes, bacterial proteins that cut the DNA at a particular short DNA sequence. If you "digest" (cut up) a big chunk of DNA with a restriction enzyme and separate the resulting smaller chunks with electrophoresis, you get a particular pattern of fragments. Kan and Dozy found that the pattern was different for sickle cell and normal DNA: the enzyme failed to cut at a spot near the sickle cell allele, resulting in a longer-than-normal fragment in the sickle cell DNA. You could, in short, reliably identify sickle cell DNA by the presence of a fragment 13,000 nucleotides long, in place of the normal 7,600-nucleotide one. It was a *length polymorphism* based on restriction fragments—RFLP. In the hallways, researchers called them "riflips."[47]

Botstein, White, Skolnick, and Davis realized that there must be thousands of such polymorphisms, spread across the genome. Their insight was to conceive of making a map of restriction fragment length polymorphisms for every chromosome. It would be like a system of mile markers over the entire genome, which gene mappers could then use to orient when pinpointing the location of specific alleles. In their now-classic paper of 1980, Botstein and colleagues issued a call to arms for a RFLP map of the genome. They outlined the procedure, provided details about the number of RFLPs that would be needed for reliable coverage, and mentioned risks and potential pitfalls. A principal advantage over human-mouse hybrids, they wrote, was that with RFLPs you did not need to understand the underlying biochemistry in order to find the gene. Unburdened by false modesty, they concluded that their technique "should provide a new horizon in human genetics," particularly in terms of preventive medicine.[48]

Riflips took polymorphism down to the DNA. Previously, a genetic marker had been a trait—either something one could see with the naked eye or else a chemical trait, a band on a gel or a spot on paper. Now a genetic marker was on the DNA itself. Once one had a RFLP, one could find the actual gene. Techniques such as gene cloning and DNA sequencing—introduced in 1975 by Frederick Sanger and then, by another method, by

Watson's protégé and colleague Walter Gilbert, working with Albert Maxam—made it possible to start from a RFLP marker and "walk" along the chromosome in either direction until one found the gene itself. It was arduous, but one could locate and analyze the actual genetic material associated with strongly hereditary disease. This led to a golden age of the germ theory of genes.

With these tools researchers rushed into the orchard of hereditary disease genes and picked the low-hanging fruit: common, devastating diseases with medical relevance, simple Mendelian heredity, and great big genes. Back in 1938 William Allan had called Duchenne muscular dystrophy a "lethal and unnecessary disease," something he would happily eliminate from the population if he could. In 1983, less than three years after the RFLP technique appeared, Robert Williamson's group at St. Mary's Hospital in London had the Duchenne gene—which codes for a protein named dystrophin—accurately mapped on the short arm of the X chromosome, and by 1987 Lou Kunkel's group at Harvard had cloned the entire massive gene, more than two million bases long and containing more than sixty coding segments spliced together to make the final message.[49] The next year, a team led by Francis Collins at the University of Michigan cloned the gene for cystic fibrosis, narrowly winning a much-publicized "race" with Williamson.[50] This was competitive, fast-paced, high-tech science. The staid pages of *Science* and *Nature* sounded like the *National Inquirer* with glassware and acronyms; the back-stabbing and rule-bending in the hunt for the cystic fibrosis gene made *The Double Helix* sound like Ozzie and Harriet. In 1983 Nancy Wexler cloned the gene that is mutated in Huntington disease—a story made touching by the fact that Wexler's father had the disease. Wexler's passionate search for the gene gave the new medical genetics the feel of a personal quest.[51] And like a soap opera, the heartbreaking story of the search for a genetic allele associated with a predisposition to breast cancer was reported serially over several years, with those who followed it anxiously awaiting the next installment. Everyone's heroine, Berkeley's Mary-Claire King, had been preparing and searching much of her adult life to find the gene, only to be beaten to the finish line by Mark Skolnick.[52]

The new techniques reinforced the conception of genes *for* disease or other traits. Genes now had a material basis: you could find one, clone it, analyze it. Wexler, whose father died of Huntington disease, had a 50 percent chance of inheriting the mutant gene. Did she have "the gene"? Wisely, she refused to state publicly whether she had even taken the test. Like Paul de Kruif's romantic microbe hunters of a century before, the gene hunters bagged the big game of disease, tracking down the causal agent and isolating it. The promise, of course, was that with the cause in hand, a treatment or even a cure would follow closely. Gene cloning therefore reinforced the germ theory of genes—the notion of a gene as a disease agent, a thing that could be isolated and either modified or eliminated.

The pinnacle of this glamorous new genetics was gene therapy—Lederberg's "algeny." Fantasies of snipping out diseased genes and replacing them with healthy ones, or even curing or repairing genes, have a long history; a certain swashbuckling style of scientist has always been attracted to the idea, even before molecular techniques suggested mechanisms for doing it. In 1970 Stanfield Rogers, a senior scientist at Oak Ridge National Laboratory, injected twin girls with a tumor virus called papillomavirus. The girls had a serious though exceedingly rare metabolic disorder, resulting from the lack of a single enzyme. Preliminary results suggested that papillomavirus produced that enzyme the girls lacked. Would infecting the girls with the tumor virus cure the disease? It was an astonishingly cavalier bit of human experimentation, carried out on the eve of genetic engineering. Rogers was censured by the biomedical community. But then, in 1979, five years after Asilomar, Martin Cline, a researcher at UCLA, proposed an experiment to insert the beta globin gene—which encodes a major subunit of hemoglobin—into patients. If production of healthy beta globin could be stimulated, a variety of blood diseases could be treated, such as sickle cell disease and thalassemia. In the approved version of the experiment, Cline was to build an elaborate construct consisting of the "healthy" human beta globin gene, bacterial operon genes, and viral genes—but he was to snip the whole thing apart with restriction enzymes before injecting it into patients. It was to be a proof of concept and a test of toxicity, not an actual therapeutic trial. But the temptation was too much for Cline; he flew overseas to conduct the trial with the live recombinant vector. He, too, was censured by the National Institutes of Health, was unable to get funding for years, and suffered irretrievable damage to his career. "I guess 'hubris' is the word that would come to mind," he said.[53]

"At what point, then, is it ethical to begin a revolutionary new treatment procedure?" asked W. French Anderson the next year. Recombinant DNA and genetic engineering techniques stood poised for application to medicine. The work was experimental in the most basic sense: little to nothing was known about dosage, methods, immediate risks, long-term risks, or benefits. Scientists found themselves revisiting the Nuremberg Trials for guidelines on human experimentation. A large bioethical apparatus was established to cope with the new biomedicine, including the federal Recombinant DNA Advisory Committee and Institutional Review Boards at every research institution to review research grant applications. In 1979 the Belmont Report established American federal policy for human subjects research. Genetic counseling, practiced for decades by physicians and a few medically oriented scientists, professionalized, with the establishment of a master's-level program at Sarah Lawrence College in 1969, expanded by Joan Marks in 1974. Asilomar and the XYY controversy were not isolated events; they were part of a Great Awakening of bioethics, set amid and triggered by the technical advances of biotechnology.[54]

Hermann Muller's dream seemed to be coming true. It was becoming technically possible to manipulate human genes and thereby simultaneously treat genetic disease and improve the gene pool. To perfect the human race. It seemed the clumsy expedients of human improvement via controlling marriage or birth would soon be unnecessary; one could go straight to the genes. Where once eugenicists had tried to cure defective populations or individuals, French Anderson could now say, "I was going to cure defective molecules."[55]

Anderson understood the lineage of the science of human perfection. In a retrospective chapter published in 1997, he traced the roots of gene therapy back to the early sixties. He identified it with Lederberg's "algeny" and linked it to Lederberg and Stanfield Rogers's conception of using viruses to introduce foreign genes into people. Anderson himself began publicly speculating on the possibilities of gene therapy as early as 1968. And the 1997 piece quoted the phage geneticist Robert Sinsheimer in 1969:

> The old eugenics would have required a continual selection for breeding of the fit, and a culling of the unfit. The new eugenics would permit in principle the conversion of the unfit to the highest genetic level. The old eugenics

> was limited to a numerical enhancement of the best of our existing gene pool. The horizons of the new eugenics are in principle boundless—for we should have the potential to create new genes and new qualities yet undreamed. . . . Indeed, this concept marks a turning point in the whole evolution of life. For the first time in all time, a living creature understands its origin and can undertake to design its future.[56]

Eugenics had meant many things over the preceding century, from tax incentives for the upper class to the exclusion of the foreign-born to the sterilization and even murder of the ill and strange. The constants in its history were the desire to control our own evolution by reducing suffering and eliminating disease, and the belief that now we finally had the knowledge to do it right.

Anderson drew a sharp distinction between somatic gene therapy and germ-line genetic enhancement. Correcting a missing or malfunctioning gene in the somatic tissues, he insisted, was ethically no different from any other medical procedure. The therapy affected only the patient and was used only to restore that individual to normal health. Attempts to enhance normal function—one common example was increasing production of growth hormone in people of short stature—or to change the gametes so that one was effectively treating not just the individual but his descendants as well were unacceptable. Since the late thirties, negative eugenics had been seen as morally superior to positive eugenics. Population improvement through preventive genetic medicine averted suffering; medicine's role was to enable people to live normal lives, not to engineer a super-race. Anderson, steeped in the literature of genetic prevention and improvement, absorbed this ethos. In 1990 he wrote, "We should not step over the line that separates treatment from enhancement."[57]

That was perhaps the last moment of moral clarity in the history of gene therapy. Later that year, Anderson and colleagues including Michael Blaese and Kenneth Culver began a gene therapy experiment in which a three-year-old girl was injected with virus engineered to ferry a missing gene into her cells, in the hope of reversing her adenosine deaminase (ADA) deficiency, an extremely rare condition that wipes out the entire immune system by disabling a single enzyme. Like so many experiments in the history of medical genetics, this trial pitted the values of science against those of medicine—and this time medicine trumped. The girl was given a new gene, but as a safety precaution she was kept on regular infusions of the adenosine deaminase enzyme the gene was supposed to produce. She and her trial-mates are alive and well, but it is not clear to what extent gene therapy is responsible. Anderson, meanwhile, underwent his own trial: a criminal trial for child molesting. He was convicted in 2006 and later sentenced to sixteen years in prison. Hubris was still a word that came to mind.[58]

Twenty years of clinical research since that first legitimate trial have produced a few modest successes and many failures. In Philadelphia in 1999, a young man named Jesse Gelsinger died from massive organ failure as a result of unforeseen consequences of gene therapy. Another trial, in France, seemed to be a success until the French girls undergoing the therapy developed leukemia, because the therapeutic gene liked to insert itself behind the promoter for a gene that led to runaway cell growth. Recently, gene therapy for beta-thalassemia has shown enough promise to prompt some to hail a comeback for gene therapy, although the predicted applications are much more modest than they once were, and here too a risk of cancer strangely parallel to the French trial has some observers worried. The cases in which a gene can be treated as a disease agent and neutralized, it seems, are few. A consensus is emerging that the germ theory of genes realistically applies only in rare and limited cases.[59]

* * *

Under Watson, Cold Spring Harbor's seedy charm gave way to moneyed, corporate polish. The chipped linoleum hallways crammed with instruments were replaced with wood-paneled, brass-railinged thoroughfares wide enough for two freezers to pass. The overgrown, brambly paths were hacked clear, and they now opened onto manicured lawns installed with statuary. DNA imagery bloomed throughout the grounds: a bell tower adorned with "a," "c," "t," "g," and a helical staircase; a fifteen-foot bronze double helix in the auditorium lobby; stylized DNA fingerprinting etched into the frosted-glass light fixtures. More than two hundred full-time researchers staffed nearly forty laboratories; at mealtimes they crowded a dining hall run by an award-winning chef, and in the evenings they gathered in an English-style pub downstairs. Thousands more scientists passed

7.2 James D. Watson at Cold Spring Harbor Laboratory, 1974, with Norton Zinder. Courtesy of Cold Spring Harbor Laboratory

through, attending the now-year-round schedule of meetings and courses. In 1998 Cold Spring Harbor Laboratory became an accredited, Ph.D.-granting university, capping its transition from unkempt scientific playground to prosperous biomedical powerhouse. There was no denying its beauty and success, though some murmured that old Jim had made Cold Spring Harbor into a place where young Jim could never have flourished.

The range of Cold Spring Harbor science expanded to include crystallography, neurobiology, and AIDS research. By the turn of the century, DNA lay at the basis of all science at Cold Spring Harbor. Indeed, to a first approximation there was no such thing as DNA science any more. All basic biomedical research had gone molecular; DNA and proteins—genes and gene products—were the foundation of biology. Increasingly, DNA became the sine qua non of biological explanation: one could not claim to understand health or disease without providing the gene and describing the action of the protein it made. It seems obvious, almost beyond question, that this must be so. Jim said Francis said that DNA is the "secret of life," and the half-century of science since seemed to confirm it beyond dispute. Watson had spent his career making sure of it.[60]

As DNA moved to the core of basic biomedical research, it became a cultural icon, simultaneously sacred and profane, an emblem of the essential true self and of our animal nature. DNA has come to stand for something's essence: a passenger car has racing in its DNA; DNA perfume has the "power of heredity"; even denim pants have genes. As Watson made a fetish object of the double helix on the grounds of Cold Spring Harbor, so has Western society done, more broadly. DNA's twinned, twining strands are now a universal icon signifying high-tech biomedicine, the way the Bohr atom became the signifier of nuclear energy. Indeed, as nuclear physics was the dominant science of the first half of the twentieth century, molecular biomedicine is the dominant science of the second half. The iconic scientist of the Cold War was Albert Einstein—wild-haired, sweetly irreverent, and almost piously concerned with the political and spiritual implications of his discipline. That of our own time is Jim Watson. His hair, too, tends to fly away, though it is much thinner than Einstein's; he too expressed ethical concerns about his science, although he soon renounced them in favor of scientific and commercial freedom; and his irreverence has been anything but sweet.[61]

As the Jim in *The Double Helix* is a comic figure, always willing to take a pratfall to make his hero Crick look good, so Watson in real life sacrificed his dignity on behalf of DNA. His fame had always been founded in part on his indiscretion. Yet later in his career he began to make remarks that offended even his most indulgent supporters. He played on the worst stereotypes of race, gender, and IQ and gave them determinist biological explanations. As the world leader of DNA, he was the natural choice to lead the new National Center for Human Genome Research in 1988, but he stepped down four years later over acidic, name-calling conflicts with NIH director Bernadine Healy. In 2000 he implied that dark-skinned people had a greater sex drive than lighter-skinned northern Europeans. In 2003 he said that "stupidity is a disease" and he would like to "help" the bottom 10 percent of the IQ bell curve by getting rid of it—a striking instance of unsympathetic Galtonian reasoning. When, in 2007, he said he was "inherently gloomy about the prospect of Africa" because "all our social policies are based on the fact that their intelligence is the same as ours—whereas all the testing says not really," the Cold Spring Harbor trustees asked for his resignation as chancellor of the Lab. He had spent nearly forty years as its top man. Most of his off-color remarks had a bit of legitimate evidence at their core—a new finding about an enzyme, or a family study suggesting a hereditary predisposition. But Watson was too willing to take evidence for a molecular mechanism and turn it into a dogmatic statement about hereditary causation. In the end, he fell prey to the same temptation to which the old Progressive eugenicists had succumbed: he let genetic determinism amplify his prejudices and biases. It is, perhaps, an occupational hazard of those who think about genes too much.

Just before Watson stepped down from Cold Spring Harbor, he made more upbeat news: a biotech company, 454 Life Sciences, had sequenced his genome. "Project Jim" took two months and cost nearly a million dollars—a lot of time and money, but orders of magnitude less than it had cost to do the first, composite genome for the human genome project. Craig Venter, who as president of Celera Corporation was the government's competitor and collaborator, published his own sequence soon after. Watson has made his complete DNA sequence publicly available. It was a bit of a publicity stunt, but Watson was, as ever, forging ahead, pushing the next stage in our evolving relationship with our own heredity.[62]

* * *

In some ways, Watson looks very much like his predecessor at Cold Spring Harbor, Charles Davenport. Both were trained as geneticists and then moved into studies of human heredity out of a desire to apply genetic knowledge to human social problems. Each often demonstrated a sophisticated, up-to-the-minute understanding of the findings of his field. Each rightly

7.3 The Hazen bell tower at Cold Spring Harbor Laboratory. It features a helical staircase and the initials of the four DNA nucleotides on each of its faces. Photo by Margot Bennett, courtesy of Cold Spring Harbor Laboratory earned his colleagues' respect but developed the reputation of being a loose cannon. The power of genetic explanations of disease, intelligence, and personality seduced both men, although when pushed, both recognized that complex traits are produced by many genes, that most genes have many different effects, and that the environment plays a large role in determining gene action. Like Watson, Davenport promoted genetics as the most powerful weapon in our armamentarium for eliminating disease and perfecting the human race for universal benefit. And like Davenport, Watson has allowed his genetic determinism to amplify his prejudices and assumptions about humanity.

Yet Watson's career is embedded in a very different social context from Davenport's. The Human Genome Project got under way in 1988; the next year the Berlin Wall came down and not long after, the Soviet Union broke up. The genome age is also the post–Cold War age. Genetic determinism and eugenic cheerleading have different meanings today than they did in the Progressive era. First and most obvious, the technological context is dramatically different. Davenport had few means at his disposal: in order to guide evolution he had to influence people's marriage and mating habits. Today, anyone has access to birth control. The moderately well off can choose from a variety of methods of regulating and engineering their own reproduction, including in vitro fertilization, preimplantation genetic diagnosis, surrogacy, and personal genome profiling. And the rich could in principle select an embryo with a genome tailored to their preferences. Thus genetic determinism today has teeth it simply did not have a century ago. For all this power, however, technology remains limiting. Direct engineering of human genes has not yet become commonplace, after four decades of promise. Beyond simple screens for a few common diseases, "designer babies" are not a significant subset of the population. This could change rapidly, and soon. But for the moment, the primary reasons biomedical genomics focuses on relief of suffering rather than human improvement are technical, not moral.

The other major difference between America in 2012 and in 1912 is that we have less of an appetite for collectivism now. Progressives believed in progress, in evolution toward perfection. In place of Davenport's message of individual sacrifice for the greater good has grown a fierce individualism.

Genetic determinism today means looking out for one's own hereditary interests, the gene pool be damned. And there is money in DNA. For-profit companies benefit from encouraging genetic determinism among the public. There is thus high short-term motivation to persuade consumers that genes are exactly what Davenport said they were: the most important factors in determining who you are and how successful you and your offspring will be. We are taking evolution into our own hands, as the early eugenicists had wished, but we are doing it with no design, no higher purpose, no public spirit. (Indeed, no real understanding of genomic evolution, either.) Our individualism immunizes us against the eugenic hysteria that resulted in *Buck v. Bell,* state sterilization laws, and the Progressive era's cruel intolerance of difference and weakness. It is the main reason that a return of Progressive-style eugenics is not a threat today. But the return of Progressive eugenics is a red herring. Most eugenics—and in particular, most medical eugenics—has not been of the coercive, rigid Progressive style. It has been more Galtonian, more flexible, more volunteerist. Eugenics in the twenty-first century has a unique character. Driven by technological power that outstrips biological understanding and guided by the principles of individual gain rather than social progress, the pursuit of human perfection becomes short-sighted and self-centered, both for better and for worse.

Epilogue

THE EUGENIC IMPULSE

IN 2008 COLD SPRING HARBOR LABORATORY, epicenter of the American Progressive-era eugenics movement, published a book titled *Davenport's Dream*. Included is a facsimile of Charles Davenport's big book of 1911, *Heredity in Relation to Eugenics*. The reissue might have been merely a welcome if quirky gesture of archival salvage but for being prefaced with nearly two hundred pages of essays by scientists, historians, and legal experts. Those essays transform the volume into a remarkable attempt to restore Davenport's reputation and reopen a discussion of eugenics for the genome age. "Charles Davenport had the best of intentions," writes Matt Ridley in the foreword, but his idealism got ahead of his knowledge and of the available technology. Elof Carlson, a student of H. J. Muller's, writes that "the role of eugenics in our time is in maximizing [hereditary] information and its availability to those who need it and minimizing the temptation to use the State as the means of enforcing eugenic ideals." The genome guru Maynard Olson writes that the publication of the human genome and the database of genetic polymorphisms known as SNPdb make possible the fulfillment of Davenport's dream. "Here," Olson writes, "is the raw material for a real science of human genetic perfection."[1]

Davenport, of course, thought *he* had the raw material for a real science of human perfection. Francis Galton's original conception of eugenics was based on the agriculturalist's subjective, holistic understanding of heredity. The advent of Mendelian genetics, it seemed to Davenport and his contemporaries, at last placed eugenics on a sound scientific footing. In the middle of the twentieth century, human-geneticists disavowed their debt to Davenport. Those who acknowledged his pioneering work in medical genetics dug a trench between themselves and Progressive-era eugenics, while others such as Victor McKusick simply rewrote the history as though it had begun in 1957. For fifty years, Davenport was geneticist non grata. In the preceding pages, we have seen that eugenic dreams did not by any means go away, but until recently, *eugenics* was a dirty word and Davenport was treated like a Nazi grandfather, an embarrassing stain on the pedigree of a now-legitimate and humane medical field. What changed? And what didn't?

The completion in 2000 of a reference sequence for the human genome marked a threshold for the Garrodian approach to human heredity that had been gathering since midcentury. This biochemical, disease-oriented approach is typically portrayed as the means by which human genetics escaped its eugenic past. Nuanced Garrodian thinking, the argument runs, prompted the shift from a simplistic approach to complex traits to a complex, multifactorial approach to simple, well-defined ones. It therefore helped geneticists reject the simple-minded determinism and ideology of the eugenics era. We have seen how biochemical genetics, combined with cell biology, enabled human-geneticists to do genetics without sex, thereby escaping the ethical problems that had dogged the field since its inception and allowing those scientists to begin the defining projects of contemporary genetic medicine: human gene mapping and genetic engineering.

Further, the medicalization of human genetics gets the credit for keeping human-genetic modification from reverting to eugenics. If genetics is medical, the advocates tell us, the science is about treating and curing disease, and therefore it can't turn into morally repugnant eugenics. Early gene therapists such as French Anderson took this line. In 1985 Anderson distinguished clearly between somatic and germ-line gene therapy, arguing that "genetic engineering should not be used for enhancement purposes" except under what he considered very precise circumstances of preventive medicine: "Regardless of how fast our technological abilities increase, there should be no attempt to manipulate, for other than therapeutic reasons, the genetic framework (i.e., the genome) of human beings." Again, in 1989: "Once we step over the line that delineates treatment from enhancement, a Pandora's box would open. On medical and ethical grounds a line should be drawn excluding any form of enhancement engineering."[2]

With astonishing speed, that line was erased with only the ghost of a trace remaining—and it is precisely the developments in molecular biomedicine that wiped the slate. Genetics without sex removed the technical boundary: at the DNA level, there is no difference between therapy and enhancement. You can upregulate growth hormone just as easily in a dwarf, a person of short stature within the normal range, or someone of average height who simply wishes to be tall. Increasingly, the distinction between somatic and germ-line gene therapy seems arbitrary, unconstructive, and anyway unenforceable.[3]

Genetic medicine has, in fact, redefined disease. By the early years of the human genome project, diseases began to be defined not by their symptoms but in terms of their genetics. In 1994 Eric Lander and Nicholas Schork described the gene hunter's technique of "stacking the deck in your favor" by narrowing the definition of a disease so that your gene of interest explains more of the variation. For instance, parsing one disease into early- and late-onset forms, or mild and severe, might clear away confounding variation and reveal a strong genetic basis. In this view, what were "really" two (or more) diseases had been conflated because of similar symptoms; genetic analysis enables one to tease them apart. Disease, of course, is a human construct. Defining it by its mechanism isn't any more *real* than defining it by its symptoms. But the advantages of conceiving disease in genetic terms are quite real indeed; not the least among them is that biochemical mechanisms present clear pharmaceutical targets. Genetic medicine is a gold mine for the drug industry. It is also a gold mine for diagnosticians. It has always seemed so. In 1912 Harvey Ernest Jordan, commenting on the place of eugenics in the medical curriculum, wrote, "Modern medicine, yielding to the demands of real progress, is becoming less a curative and more a preventive science." Eighty years later, Lee Hood, one of the pioneers of DNA sequencing, wrote, "The diagnosis of disease-predisposing genes will alter the basic practice of medicine in the twenty-first century. . . . Medicine will move from a reactive mode (curing patients already sick) to a preventive mode (keeping people well)." Genome medicine, then, is realizing the pipe dreams of medically oriented eugenicists in the Progressive era.[4]

Then as now, preventive medicine seemed unassailably benevolent. But as biomedical researchers increasingly define disease as well as treat it, the public inadvertently cedes management of individual health to a for-profit healthcare industry. This industry is founded on the principles and vision whose development I have traced in the preceding pages. In 2009 Johns Hopkins Medical School implemented a new curriculum called Genes to Society, anchored in the Garrodian approach championed by Barton Childs. Hopkins has made Genes to Society the "brand"—the school's term—of its increasingly aggressive efforts to export its approach and acquire teaching hospitals out of state and internationally. Garrodian medicine is big business. One of its most distinctive features is its expansion of the medical clientele: doctors no longer treat only sick people. When David Nichols, vice dean for education at Johns Hopkins School of Medicine, discussed Genes to Society, he stressed biomedicine's growing emphasis on prevention and detection of disease in its "latent phase." Latent disease has no symptoms; it is detectable only by a high-tech biomedical researcher. When I asked Nichols who, then, counted as a patient, he said, "We are all patients now." This, of course, dramatically expands the constituency for genetic and other medical tests, as well as prophylactic, possibly lifelong, prescriptions of pharmaceuticals.[5]

In this climate, the moral clarity of French Anderson—once the consummate gene cowboy—seems naïve. Drawing a line between prevention and enhancement would require innumerable arbitrary decisions about the medical necessity of dietary choices, physical appearance, and psychological hardship. How can we distinguish between preventing heart disease and minimizing the consequences of a poor diet? Would reducing the chances of alcohol abuse be therapy or enhancement? In the end, no logical friction can slow the slide from prevention to enhancement—or from the individual to the population. The genome project erased Anderson's line and completed the blending of Galtonian and Garrodian approaches.

"There is every difference between the goal of individual eugenics and Davenport's goal," Ridley writes in his introduction to *Davenport's Dream*. "One aims for individual happiness with no thought to the future of the human race; the other aims to improve the race at the expense of individual happiness." But that's not quite right. We have seen that "Davenport's goal" (a euphemism for Progressive eugenics) included individual happiness—even if it did subordinate it to that of the future population. Eugenicists from Galton forward sought to make individuals happy *by way of* the population. It was a matter of accessibility. In the nineteenth and early twentieth centuries, the population seemed more amenable to engineering, more malleable. Changes in technology, scientific and medical knowledge, and social values slowly made individuals, cells, proteins, genes, and finally nucleotides accessible to analysis and available to modification. So far, much of the genomic revolution remains at the level of promises, as far as the public is concerned. Genetically enhanced in vitro babies remain science fiction. My argument has been that these promises have been consistent throughout the history of medical genetics, and that promises have power.

Even more striking, Ridley here implies that the aim of individual happiness *is* twenty-first-century eugenics. In other words, the ideological membrane between relief of suffering and human improvement has dissolved. I have shown that relief of suffering has always been a part of eugenics. Now Ridley suggests that it has *become* eugenics. That's a remarkable rhetorical move, but it's disingenuous. Today's eugenicists by no means ignore the future of the race. *Davenport's Dream* was only the most sober of a raft of pro-eugenics books published in the first decade of the century. The literature has a millennial feel, with such titles as *Redesigning Humans, Radical Evolution, Enhancing Evolution, More than Human,* and *The Price of Perfection.* Biomedicine grades almost insensibly into cultish "transhumanism." Twenty-first-century eugenicists, while grounded in egotistical individualism, cannot resist the allure of taking control of our own evolution, of engineering not just our development but our future. They gleefully transgress the last sacred moral genetic boundary, solemnly etched by French Anderson. Why not, they ask, design children according to our whims and tastes? Genetic medicine, they tell us, is increasingly enabling us to engineer ourselves. We can choose our traits, pick our predispositions, prevent disease and weakness. Many of these authors, like the scientists in *Davenport's Dream,* use the word *eugenics* once again. Often it is called *individual eugenics* or *liberal eugenics.*[6]

The genetic knowledge and molecular technology that made the individual accessible to engineering are also making the population accessible. Individual and population were once treated as parts of a single science of natural history. Darwin's *Origin of Species* began to dissociate heredity, development, and evolution as distinct scientific problems. That dissection enabled genetics, embryology and developmental biology, and evolutionary biology to differentiate and mature as scientific fields. But during the twentieth century they have gradually merged again, into a single science of growth and form. Molecular genetics and genomics have united developmental biology, the science of changes in individuals, with evolutionary biology, the science of changes in populations. Precisely through the techniques of changing individuals, scientists are now contemplating changing populations—"the self-direction of human evolution," as the old eugenics poster put it. For example, in *Enhancing Evolution,* John Harris proposes "both the wisdom and the necessity of intervening in what has been called the natural lottery of life, to improve things by taking control of evolution and our future development to the point, and indeed beyond the point, where we humans will have changed, perhaps into a new and certainly into a better species altogether." The genome project, then, has enabled eugenicists to come out of the closet. The fantasy of controlling our own evolution is alive and well.[7]

* * *

The return of eugenic language to discussions of the promise of medical genetics should not surprise any reader who has made it this far. Throughout this book, I have treated eugenics not as a historical aberration but as a constant, continuous impulse running through the recent history on health and human heredity. This has been no scholarly contrivance; I have merely adopted the inclusive view of eugenics employed by medical geneticists throughout the twentieth century. Galton sought "not only the restoration of the average worth" of the human race, debased as he saw it by disease and mental degeneration, "but to raise it higher still." That "restoration" included the elimination of hereditary disease and disability, both physical and mental, and the strengthening of the constitution against infectious disease. Raising it higher still meant improving humankind physically, intellectually, morally, and spiritually. It was to be done, he believed, "under existing conditions of law and sentiment." Galton understood eugenics to be something that everyone capable of rational thought and possessed of a humanitarian, communitarian spirit would embrace willingly.[8]

This impulse to improve human health and happiness with genetics underlies the promises of genetic medicine. A longer life, less disease, greater intelligence, and a better adjustment to the conditions of society have always been the goals of hereditary enthusiasts. To say that our eugenics is better than that of an earlier time is only to say that eugenics is best when it is adapted to current social conditions; it amounts to a truism. In the Progressive era, Ridley's individualist eugenics would have seemed appallingly and dangerously egotistical. To identify contemporary genetic medicine with the eugenic impulse does not necessarily make high-tech healthcare insidious. Conversely, it does complicate the notion that eugenics is necessarily a social evil.

The eugenic impulse arises whenever the humanitarian desire for happiness and social improvement combines with an emphasis on heredity as the essence of human nature. It is the dream of control, of engineering ourselves, of not leaving our future up to cruel fate. This impulse is noble in spirit but, unleavened by an equal impulse to improve the conditions of life, it is deceptive and ultimately impoverishing. Human happiness is overdetermined—more than one set of causes can provide a complete account of it. Even a full molecular explanation of health or intelligence or personality would not preclude an equally complete explanation in terms of upbringing and training. Heredity trumps environment by collective decision, not natural necessity. We choose to explain human nature in terms of heredity because it offers technological solutions that, challenging and expensive as they may be, are ultimately easier and sexier than social solutions. The greatest risk of hereditary determinism may be not the results it produces but the alternatives to which it blinds us. It obscures the power of diversity, the beauty of chance, and the virtues of tolerance, by creating an illusion of perfectibility.

NOTES

1. The Galton-Garrod Society

1. McKusick, "60-Year Tale," 4–5; McKusick interview.
2. The canonical biography of Galton is Pearson's admiring and detailed *Life, Letters, and Labours.* Kevles, *In the Name of Eugenics,* chapter 1, offers a skeptical sketch that contains many insights. In recent years, three book-length biographies of Galton have appeared: Bulmer, *Francis Galton*; Brookes, *Extreme Measures*; and Gillham, *A Life of Sir Francis Galton.* Much of what follows is drawn from these. Also, numerous recent articles examine Galton's contributions to specific fields, for example, Sweeney, *"Fighting for the Good Cause"*; Walker, "Putting Method First"; Moss, "Introductory Note to a Classic Article"; Waller, "Gentlemanly Men of Science"; Aronson, "Francis Galton and the Invention of Terms for Quantiles"; Burbridge, "Francis Galton on Twins, Heredity, and Social Class"; and Nagy, "Galton and Proficiency Testing."
3. Pearson, *Life, Letters, and Labours,* 1, 144.
4. Nearly all of Galton's published works, as well as criticism, reviews, and other ancillary material, are collected in facsimile form at the superb resource http:/galton.org. Romanes, "Human Faculty," 97.
5. Galton, "Relative Sensitivity at the Nape of the Neck"; Bulmer, *Francis Galton,* 250.
6. Galton, "Statistical Inquiries."
7. Galton, *Natural Inheritance,* chapters 6–7.
8. Gillham, *A Life of Sir Francis Galton,* chapter 18. On William Farr and the statist origins of statistics, see Eyler, *Victorian Social Medicine.* On Pearson, see Porter, *Karl Pearson.*
9. Darwin, "Provisional Hypothesis of Pangenesis"; Galton, "Theory of Heredity," 330. See also James Schwartz's lively account, *In Pursuit of the Gene,* chapter 7.
10. Galton, "History of Twins," 566, 576; Galton, "Hereditary Improvement," 116.

11. Galton, *Hereditary Genius*, 375–76.
12. Galton, *Hereditary Genius*, 376; Galton, "Hereditary Talent and Character," 165; Galton, "Possible Improvement of the Human Breed" (1901), 665. On the costs of utopias to liberty, see Wegner, *Imaginary Communities*.
13. Galton, "Hereditary Talent and Character," 165; Galton, *Inquiries into Human Faculty and its Development*, 17.
14. Galton, "Possible Improvement of the Human Breed" (1901), 664. Galton to Bateson, June 12, 1904, quoted in Pearson, *Life, Letters, and Labours*, 220–21.
15. Galton, "Possible Improvement of the Human Breed" (1909), 24; this is a reprint of the 1901 essay of the same title, used for its slightly more felicitous wording.
16. Various, "Discussion of the Advisability of the Registration of Tuberculosis."
17. Galton, "Eugenics and the Jew."
18. Bearn, *Archibald Garrod and the Individuality of Man*.
19. Brian, *Pisse-Prophet*. See also Stolberg, "Decline of Uroscopy."
20. Wolkow and Baumann, "Ueber das Wesen der Alkaptonurie"; Bearn and Miller, "Garrod and the Concept of Inborn Errors," 320.
21. Garrod, "About Alkaptonuria."
22. Ibid., 1485.
23. On Bateson, see Harper, "William Bateson." The idea of Bateson as conservative was put forward by Coleman, "Bateson and Chromosomes"; developed by MacKenzie and Barnes, "Scientific Judgment"; and refuted by Cock, "Bateson's Eventual Acceptance of Chromosome Theory." "Never attempted to conceal": Haldane, "William Bateson," 137. On W. K. Brooks and the Chesapeake marine laboratory, see Maienschein, *Transforming Traditions in American Biology*, chapters 1–2.
24. Gillham, "Galton and the Birth of Eugenics," 94–95; Weldon, "Study of Animal Variation."
25. Correns, "Mendel's Regel"; de Vries, "Sur la loi de disjonction"; de Vries, *Die Mutationstheorie*; Fisher, "Has Mendel's Work Been Rediscovered?"; Monaghan and Corcos, "Origins of the Mendelian Laws"; Olby, *Origins of Mendelism*, chapter 6; Schwartz, *In Pursuit of the Gene*, chapter 6.
26. Bateson, "Problems of Heredity."
27. Bateson, *Mendel's Principles of Heredity*, 27; Bateson and Saunders, "Facts of Heredity," 3; Bateson, *William Bateson*, 93.
28. Bateson and Punnett, "Inter-Relations of Genetic Factors," 3; Bateson, Saunders, and Punnett, "Physiology of Heredity."
29. Bearn and Miller, "Garrod and the Concept of Inborn Errors," 321–25.
30. Garrod, "Incidence of Alkaptonuria."
31. Ibid., 1616.
32. Ibid.
33. Ibid., 1618.
34. Ibid.
35. Ibid., 1620.
36. Ibid.

37. Garrod, *Inborn Errors of Metabolism*; Garrod, Scriver, and Childs, *Garrod's Inborn Factors in Disease*, 9.
38. Kohler, *Lords of the Fly*, 47.
39. Bearn, "Archibald Edward Garrod"; Kohler, *Lords of the Fly*, 47; Bateson, *William Bateson*. See also Coleman, "Bateson and Chromosomes"; Cock, "Bateson's Eventual Acceptance of Chromosome Theory"; Cock, "Bateson's Two Toronto Addresses"; Harper, "William Bateson."
40. Ackerknecht, "Diathesis"; Burgio, "Diathesis and Predisposition"; Mendelsohn, "Medicine and the Making of Bodily Inequality."
41. Garrod, Scriver, and Childs, *Garrod's Inborn Factors in Disease*, 62, 117.
42. Ibid., 12; Cannon, *Wisdom of the Body*.
43. Hogben, *Nature and Nurture*, 57–58; Muller, "Progress and Prospects in Human Genetics," 8; Haldane, *Biochemistry of Genetics*, 124.
44. Comfort, "When Your Sources Talk Back."

2. Fisher's Quest

1. Fisher to Margaret Hazard Fisher, December 27, 1904, f. 1904, box 2, Irving Fisher Papers, Yale University Library and Archives (hereafter Fisher Papers).
2. Kimmelman, "American Breeders' Association"; Kimmelman, "Progressive Era Discipline"; Paul and Kimmelman, "Mendel in America."
3. Allen, *Irving Fisher*, 8–10, 36–37.
4. Fisher to William Greenleaf Eliot, Jr., December 1903, f. 1898–1903, box 2, Fisher Papers; Allen, *Irving Fisher*, 83–87. Quoted passage is from Fisher to Herbert Fisher, 1903, quoted ibid., 87.
5. Rosenberg, "Piety and Social Action," 116–20; Whorton, *Nature Cures*, chapter 4, esp. 85; Smith, *Clean*, 294; Browne, *Charles Darwin*, 2: 232–38. On the religious and spiritual side of eugenics, see also ff. "Sermon Contest," box 12, American Eugenics Society Papers, American Philosophical Society Library; Rosen, *Preaching Eugenics*, 4. For development of the notion of purity and eugenics, see Cooke, "False Gods and Red-Blooded Women."
6. Weiss, *Race Hygiene and National Efficiency*, 19. On the Germans' technocratic approach to eugenics, see Weiss, "Race Hygiene Movement in Germany."
7. Fisher to Margaret Hazard Fisher, January 7, 1905, f. 1905, box 2, Fisher Papers; Schwarz, *John Harvey Kellogg*, chapters 1–3. On Trall, see Whorton, *Nature Cures*, 88–94.
8. IF to MHF, January 1, 1905; Battle Creek Sanitarium menu, January 13, 1906, f. 1905, box 2, Fisher Papers.
9. Kellogg, *Plain Facts for Old and Young*, 291.
10. Schwarz, *John Harvey Kellogg*, 220–29. On Fisher's nutritional experiments, see ff. 1905–1906, box 2, Fisher Papers; quotation about Taft from IF to MHF, April 24, 1906, f. 1906 Apr–Dec.
11. "Wonderful potencies": quoted in Castle, "Beginnings of Mendelism in America," 62. Kimmelman, "American Breeders' Association," 184. On the role

of the land grant universities, see Rossiter, "Organization of the Agricultural Sciences."

12. Pernick, *Black Stork*, 44; Davenport, *Eugenics.*
13. Thalassophilia: Allen, "Eugenics Record Office," 244–45; Kevles, *In the Name of Eugenics*, 49, chapter 3. On Davenport's life, career, and reputation, see MacDowell, "Charles Benedict Davenport"; Hiltzik, "Brooklyn Institute of Arts and Sciences' Biological Laboratory," chapters 3, 4; Witkowski, "Charles Benedict Davenport."
14. On the Statione Zoologica and the Marine Biological Laboratory, see Groeben et al., "Naples Zoological Station and the Marine Biological Laboratory"; Benson, "Naples Stazione Zoologica"; Maienschein, *Defining Biology*; Maienschein, *100 Years of Exploring Life*; Maienschein, *Transforming Traditions in American Biology*; Maienschein, " 'It's a Long Way from Amphioxus.' "
15. MacDowell, "Charles Benedict Davenport," 20. An example of Davenport's early embrace of Mendelism: Davenport, "Mendel's Law of Dichotomy in Hybrids."
16. Billings to Davenport, February 11, 1904, "CSH Beginnings—Correspondence 1904," Charles Davenport Collection, American Philosophical Society Library.
17. CBD to WHW, April 14, 1930, f. D'Auncy-Davenport, box 11, Welch Papers, Alan Chesney Archives, Johns Hopkins University; MacDowell, "Charles Benedict Davenport." On Bell's interest in eugenics, see Stansfield, "Bell Family Legacies"; Greenwald, "Alexander Graham Bell through the Lens of Eugenics."
18. Allen, "Eugenics Record Office." Hiltzik, "The Brooklyn Institute of Arts and Sciences' Biological Laboratory," chap. 5. On Laughlin, see Wilson, "Harry Laughlin's Eugenic Crusade." Kevles, *In the Name of Eugenics*, chap. 7.
19. Bix, "Experiences and Voices of Eugenics Field-Workers."
20. Davenport, "Trait Book."
21. Ibid., 2; Davenport, "Annual Report of the Director."
22. Meyer to Davenport, May 2, 1913; Laughlin to Meyer, December 18, 1913; Meyer to Laughlin, December 22, 1913, III/72/8, Adolf Meyer papers, Alan Mason Chesney Archives, Johns Hopkins University (hereafter Meyer Papers).
23. New York (State) Board of Charities, *Field Work Manual*, 8–22. Susan Lindee initiated the conversation about contemporary biomedicine seeing all disease as ultimately genetic: Lindee, "Genetic Disease in the 1960s."
24. Kevles, *In the Name of Eugenics*, 122; Spiro, "Nordic vs. Anti-Nordic"; Pickens, *Eugenics and the Progressives*, esp. chapters 3, 4.
25. Deborah Kamrat-Lang has looked at medical language in American eugenics—in some ways, the converse of what I am examining here; see Kamrat-Lang, "Healing Society." On European medical genetics, see Schneider, "Eugenics Movement in France"; Gaudillière, "Mendelism and Medicine"; Mendelsohn, "Medicine and the Making of Bodily Inequality"; Wilson, "Confronting 'Hereditary' Disease," 25–30.
26. Wiebe, *Search for Order*, esp. chapters 6–7; Rodgers, "In Search of Progressivism," 123–27; Glickman and Diner, *Very Different Age*. A moving and readable case study of an impoverished neighborhood in this period is Anbinder, *Five Points*.

27. Burnham, *How Superstition Won and Science Lost.*
28. Bernard, *Introduction to the Study of Experimental Medicine,* 66; Barker, Foreword to *Eugenics,* ix. On Loeb, see Pauly, *Controlling Life.*
29. Rodgers, "In Search of Progressivism," 113–15.
30. Norton, "Economic Advisability of a National Department of Health," 519.
31. Schieffelin, "Work of the Committee of One Hundred," 77.
32. Fisher to WG Eliot, Jr., September 26, 1906, Fisher Papers.
33. Fisher to WG Eliot, Jr., August 15, 1906, Fisher Papers; Rosen, "Committee of One Hundred"; Schieffelin, "Work of the Committee of One Hundred, 77–79, n. 1."
34. Stern, *Eugenic Nation,* chapter 4; TR to IF, May 8, 1907, f. 1907, box 2, Fisher Papers.
35. All quotations from Fisher, D. C. National Conservation Commission, and Health Committee of One Hundred, *Bulletin 30.* Human life at $1,700: 1; partly inherited and partly acquired . . . haphazard selection: 49; prolongs the lives of the unfit: 79; health ideals maintained: 100–101.
36. Harden, *Inventing the NIH,* 35–39.
37. Emily F. Robbins to W. G. Eliot, Jr., February 6, 1914, f. 1914, box 3, Fisher Papers; Knopf, "Federal Department of Health"; "Scientific Notes and News"; Harden, *Inventing the NIH, 35–39.*
38. Sanger, *Pivot of Civilization,* 59–60.
39. Meyer moves to Hopkins: Lamb, "Adolf Meyer," chapter 1, p. 48. Eugenics Section committees: "American Breeders' Association and Its Eugenics Section," pamphlet dated January 1, 1912, III/72/6, Meyer Papers. Plötz and Rüdin: Meyer to William H. Hays, March 25, 1909, III/72/1, Meyer Papers.
40. Gospel of Germs: Tomes, *Gospel of Germs.* Many persons interested: handwritten note on Eugenics Education Society flyer, undated but early 1909, II/232/1, Meyer Papers.
41. Jordan, "Place of Eugenics," 396, 398.
42. Davenport, "Eugenics and the Physician": "peculiar," 1195; "great loss," 1198; "crazy policy," 1198.
43. AMMBA: Schwarz, *John Harvey Kellogg, M.D.,* 24–26. On the second Race Betterment congress, see Stern, *Eugenic Nation,* chapter 1.
44. Robbins, "Proceedings of the First National Conference on Race Betterment," 449.
45. Ibid., 311.
46. Cooke, "Limits of Heredity"; Robbins, "Proceedings of the First National Conference on Race Betterment," 504.
47. Robbins, "Proceedings of the First National Conference on Race Betterment," 114, 230.
48. Stern, *Eugenic Nation,* chapter 1, esp. 53.
49. "Gospel of individual health": James D. Lennehan, Secretary, LEI, to Adolf Meyer, August 15, 1916, III/82/1, Meyer Papers. Fisher to Welch, December 8, 1920; Fisk to Welch, December 9, 1920, f. Life Extension Institute, box 11, William H. Welch

Papers, Chesney Archives. See also Hirshbein, "Masculinity, Work, and the Fountain of Youth."

50. Fisher, Fisk, and Life Extension Institute, *How to Live*, 176. On country life, see Kimmelman, "American Breeders' Association," 186–87.
51. Fisher, Fisk, and Life Extension Institute, *How to Live*, 157, 164.
52. Ibid., 167.
53. Paton, "Medicine's Opportunity," 329.
54. "Interim Committee of International Eugenics Congress," 65. On Harry Olson, see "Blind to a Nightmare."
55. Fisher to Members of the Committee of 100, F. 1923, box 5, Fisher Papers.
56. Pace McCabe and McCabe, "Are We Entering a 'Perfect Storm'?"

3. A Germ Theory of Genes

1. Quoted in Snyder, "Principles of Gene Distribution," 824.
2. Scandinavian eugenics laws: Roll-Hansen, "Geneticists and the Eugenics Movement in Scandinavia"; Drouard, "Concerning Eugenics in Scandinavia"; Broberg and Roll-Hansen, *Eugenics and the Welfare State*; Koch, "Eugenic Sterilisation in Scandinavia." On fitter-families and better-babies contests, see Selden, *Inheriting Shame*. Laughlin's survey: Laughlin, "Studies in Eugenics and Heredity," 60–63.
3. Barker, "Heredity in the Clinic," 604.
4. See, e.g., Newman, "Mental and Physical Traits of Identical Twins Reared Apart."
5. Guyer, "Note on the Accessory Chromosomes of Man"; Guyer, "Germinal Background of Somatic Modifications"; Guyer, "Internal Secretions and Human Well-Being"; Guyer and Smith, "Transmission of Eye-Defects."
6. Little, "Opportunities for Research," 534.
7. Allen, "Old Wine in New Bottles."
8. Jennings, "Raymond Pearl"; Allen, "Old Wine in New Bottles," 233–34.
9. Pearl, "Breeding Better Men," 9818.
10. Allen, "Old Wine in New Bottles," 240–44.
11. Tracy, "Draper and American Constitutional Medicine"; Tracy, "Evolving Science of Man."
12. Draper, "Relationship of Human Constitution to Disease."
13. Faber, *Nosography in Modern Internal Medicine*, 187.
14. Tracy, "Draper and American Constitutional Medicine," 56.
15. Draper, "Biological Philosophy and Medicine," 123; Garrod, Scriver, and Childs, *Garrod's Inborn Factors in Disease*, 50. On Draper and the Galton Society, see Tracy, "Draper and American Constitutional Medicine," 76–78. On Sheldon's corruption of constitutionalist holism, see Tracy, "Evolving Science of Man," 176–79.
16. Pearl et al., "Studies on Constitution," 11.
17. For a contemporary example of the constitutionalist-genomic approach to longevity, see the numerous public talks and appearances of the eccentric

gerontologist Aubrey de Grey and the SENS Foundation (http://www.sens.org/users/aubrey-de-grey).

18. Unmarked file, box 433591630 "Constitution Clinic Miscellaneous Recs," Pearl Papers, Chesney Archives, Johns Hopkins.
19. Pearl, "Biology of Superiority," 260. On Pearl's criticism of Eugenics, see Allen, "Old Wine in New Bottles," 240–44.
20. "Benefit the women of Baltimore": Jean Keyser to Meyer, April 28, 1933, III/62/3, Baltimore Birth Control Clinic, Meyer Papers. On the Birth Control Federation and the merger with Planned Parenthood, see Fee, *Disease and Discovery*, 209.
21. Andrew Mendelsohn elegantly traces the interdependent dynamic of early medical genetics and medicine in "Medicine and the Making of Bodily Inequality." On holism in biomedicine, see Lawrence and Weisz, *Greater than the Parts.*
22. Herndon, "William Allan"; Blodgett, "North Carolina Medical College."
23. See Allan, "Effect of Emetine."
24. Opitz, "Biographical Note—Laurence H. Snyder," 447.
25. Pauline Mazumdar, "Two Models for Human Genetics," has called the blood grouping laboratory the "Fly Room" of the human species, by analogy with Thomas Hunt Morgan's famous Columbia University *Drosophila* laboratory. On serology in World War I, see Kendrick, *Blood Program in World War II*, chapter 1. Quotations from "Minutes of the Executive Session of the Second International Congress of Eugenics," *Eugenical News* 7 (1922): 65–66.
26. Early in his career, Hirszfeld went by the Germanized version of his name, Ludvig Hirschfeld. For consistency, throughout this discussion I use the Polish transliteration he later adopted. Mazumdar, "Two Models for Human Genetics"; Keating, "Holistic Bacteriology."
27. Mazumdar, "Two Models for Human Genetics," 628; Snyder, "Human Blood Groups: Their Inheritance and Racial Significance," 233; Gould, *Mismeasure of Man*, chapter 2.
28. "On the Permanent Commission."
29. Snyder, *Blood Grouping.*
30. Allan, "Eugenic Significance of Retinitis Pigmentosa."
31. Snyder, "Linkage in Man."
32. Fox, "Relationship between Chemical Constitution and Taste"; Snyder, "Inherited Taste Deficiency"; Blakeslee, "Genetics of Sensory Thresholds."
33. See, for example, Snyder and Curtis, "An Inherited 'Hollow Chest' "; Snyder, Baxter, and Knisely, "Studies in Human Inheritance."
34. For biographical information on Macklin, see Mehler, "Macklin, Madge Thurlow."
35. McLaren, *Our Own Master Race, 130–31.*
36. Macklin, "Medical Genetics: An Essential Part," 297; Allan, "Preventing Hereditary Diseases," 231.
37. On the debate over how responsible was the germ theory for the decline in infectious disease, see Szreter, "Importance of Social Intervention"; McKeown, *Modern Rise of Population.* "Let him recall": Murlin, "Science and Culture," 82.

38. "Increasingly important": Macklin, "'Medical Genetics': A Necessity," 485. "Remarkable shift": Allan, "Relationship of Eugenics to Public Health," 74. "Our most serious problems": Allan, "Medicine's Need of Eugenics," 417.
39. Macklin and Bowman, "Inheritance of Peroneal Atrophy." "Continence, contraception and even sterilization": Macklin, "Hereditary Abnormalities of the Eye, I," 1342. "It is at once obvious": Macklin, "Hereditary Abnormalities of the Eye, III," 423.
40. Macklin, "Should the Teaching of Genetics Have a Place?" 372.
41. Snyder, "Study of Human Heredity," 541.
42. Macklin, "Should the Teaching of Genetics Have a Place?" 371. Diane Paul quotes the biologist Herbert Spencer Jennings making a similar analogy in Paul, "Genes and Contagious Disease," n. 13.
43. Macklin coined *medical genetics* more or less simultaneously here: "Should the Teaching of Genetics Have a Place?"; "'Medical Genetics': A Necessity"; "Medical Genetics: An Essential Part." Lusty infant: Macklin, "Teaching of Inheritance of Disease," 1335. "Although we cannot speak": Macklin, "'Medical Genetics': A Necessity," 486. "It is therefore of advantage": Snyder, "Genetics in Medicine," 705.
44. Macklin, "Medical Genetics: An Essential Part," 292.
45. Macklin, "'Medical Genetics': A Necessity," 486.
46. Snyder, "Genetics in Medicine"; Fangerau, "'Baur-Fischer-Lenz' in 1921–1940 Critical Book Reviews"; Fangerau and Muller, "Standard Textbook on Racial Hygiene"; Proctor, *Racial Hygiene*, 204–9.
47. Snyder, *Principles of Heredity*.
48. Snyder, "Genetics in Medicine," 707.
49. Snyder, *Principles of Heredity*, chapter 29 (400–411); "any more serious than appendicitis," 407; "we hold these truths," 411.
50. A smattering of examples: "Sterilization of Mental Defectives"; "Danger of Medicine Producing Racial Degeneration"; Weiss and Ellis, "Rational Treatment of Arterial Hypertension"; "Indications for Abortion and the Law"; Yater and Mollari, "Pathology of Sickle-Cell Anemia"; "Eugenic Problem." Quotation is from "Mongolian Idiocy," 684. Thanks to Peter Lippman, whose tireless efforts as an undergraduate summer research assistant one year uncovered hundreds of such references.

4. The Heredity Clinics

1. Allen, "Eugenics Record Office," 250–54.
2. Wilson, "Pedigree Charts."
3. Oliver headed up one of the heredity clinics described in this chapter. Glass published extensively on human problems; see chapter 6. Carlson, Muller's last student, has had a second career as a historian of genetics and eugenics. See, for example, *Genes, Radiation, and Society* and *The Unfit*.

4. Newman, Freeman, and Holzinger, *Twins*; Gershon, "Historical Context"; Rainer, "Franz Kallmann's Views on Eugenics"; Kolb and Roizin, *First Psychiatric Institute*; Mildenberger, "On the Track of 'Scientific Pursuit.'"
5. Charles Dollard interview with Frederick Osborn, February 6, 1939, f. 11, "Allan, William," III.A.1 A–B, box 1, Carnegie Corporation of New York Records, Rare Book and Manuscript Library, Columbia University Libraries, New York.
6. Osborn, "History of the American Eugenics Society," 118.
7. Rising sterilizations: Reilly, *Surgical Solution*, chapter 6 and p. 97, table 4. Family Research Bureau: "36th and 37th Annual Reports of the Michigan Academy of Science, Arts and Letters 1935, pp. 1–113," typescript, f. Institute of Human Biology—Heredity Clinic—History (Timeline and Correspondence), box 1, Department of Human Genetics Papers, Bentley Historical Library, University of Michigan (hereafter DHG). Sterilization and birth control: Kluchin, *Fit to Be Tied*, chapter 1; Schoen, "Fighting for Child Health," 90.
8. Pioneer Fund: Tucker, *Funding of Scientific Racism*, chapter 2. Wilson, "Pedigree Charts as Tools," 183. On Goethe, see Stern, *Eugenic Nation*, chapter 4.
9. "What is in my mind": Allan to Osborn, June 18, 1936, box N 307-1, f. 1936–1939, William Allan Papers, Coy C. Carpenter Library, Wake Forest University (hereafter Allan Papers); Osborn demurred: Osborn to Allan, July 14, 1936, box N 307–1, f. 1936–1939, Allan Papers. Wake Forest a proud achievement: Osborn, "History of the American Eugenics Society," 120–21.
10. Allan to Osborn, June 21, 1937; Osborn to Allan, July 6, 1937, f. 1936–1939, box N307-1, Allan Papers.
11. Osborn to Allan, July 6. 1937, f. 1936–1939, box N307-1, Allan Papers; Osborn, *Preface to Eugenics*, 29.
12. Osborn to Allan, April 21, 1938, f. 1936–1939, box N307-1, Allan Papers. Both Barry Mehler, "History of the American Eugenics Society," 120–27, and William H. Tucker, *Funding of Scientific Racism*, 50, have noted Osborn's support for negative eugenics and complicated his reputation as a moderate. Neither, however, explored the history of the heredity clinics that Osborn helped instigate.
13. Allan, "Relationship of Eugenics to Public Health," 74.
14. Allan to Osborn, December 5, 1938; Osborn to Allan, January 27, 1939, f. 1936–1939, box N307-1, Allan Papers.
15. Allan to Osborn, January 24, 1939; Osborn to Allan, February 16, 1939; Allan to Osborn February 22, 1939, f. 1936–1939, box N307–1, Allan Papers.
16. Schoen, *Choice and Coercion*, 32–38.
17. Allan to Osborn March 8, 1940, f. F121, "Allan, William," subject file, WFU.
18. Meads, Bowman Gray School of Medicine, and North Carolina Baptist Hospital, *Miracle on Hawthorne Hill*.
19. [Anonymous], "Herndon"; Allan to Carpenter, September 20, 1940, f. F121, "Allan, William," subject file, WFU.
20. Allan to Osborn, March 20, 1940, f. 1940, box N307-1, Allan Papers; Snyder, *Medical Genetics*.

21. Evans, "Lee Raymond Dice." "A Proposed Laboratory for the Study of Human Heredity in Michigan," *Michigan Academy of Arts, Science, and Letters Annual Report*, 1937, 84–87, f. Reprints, box 6, Dice Papers.
22. Opitz, "Obituary: Charles W. Cotterman," 149.
23. Crow and Denniston, "In Memory of Charles W. Cotterman," 903.
24. "Report to Dean Yoakum on Research Grant R-108 (L. R. Dice) 1940–1941," f. Inst of Hum Biol—Corresp and Administrative Material—1941–1942, box 1, DHG. Allan to Snyder, July 12, 1941, f. 1941, box N307-1, Allan Papers.
25. "Interested in what you are doing": Keppel to Dice, August 1, 1941, f. IHB 1946, Dice Papers. Herndon arrived in August: Herndon to Allan, August 20, 1941, f. 1941, box N307-1, Allan Papers.
26. Herndon to Allan, September 17, 1941, f. 1941, box N307-1, Allan Papers. I concentrate here on the Wake Forest and Michigan programs; for more on the Dight Institute, see Stern, *Telling Genes*.
27. "Drive me dippy": Herndon to Allan, October 30. 1941, f. 1941; "Ruin the academic outlook": Allan to Herndon, November 11, 1941, f. 1941, box N307-1, Allan Papers. Flyer dated December 1, 1941, f. 1941, box N307-1, Allan Papers; "To the Staff of the University Hospital," f. Institute of Human Biology—Heredity Clinic—History (timeline and correspondence), box 1, DHG. The original quotation reads, "The mills of God grind slowly, but they grind exceedingly small"—An apt characterization of the fine-grained genetic style of Cotterman, although why Herndon referred to the Michigan group as Gaelic is obscure.
28. Herndon to Allan, December 31, 1941, f. 1941; Herndon to Allan, March 16, 1942, f. 1942–1 Jan–May, box N307-1, Allan Papers.
29. "Report of Department of Medical Genetics May 1, 1943 to May 1, 1944," f. Carnegie Corporation, box 7, Allan Papers; Cotterman to Kuntz, October 2, 1944, f. Inst of Hum Biol—Corresp and Administrative Material—1943–1944, box 1, DHG.
30. Dice's publicity efforts: Paul M. Chandler, "U. of M. Heredity Clinic Seeks to Improve Human Breed," *Flint Journal*, August 14, 1943; "Heredity Clinic at U. of M. Aims to Better Human Race," *Pontiac Press*, August 14, 1943, both in f. Inst of Hum Biol—Corresp and Administrative Material—1943–1944, box 1, DHG; Postle and Postle, "Whose Little Girl Are You?" The Postles consulted Dice through their writing, including sending him drafts of the article, initially titled, "Human heredity clinic," f. 1947–1948, Bbx 1, DHG. Anderson as possible replacement for Herndon: Dice to Roy [*sic*: Ray] C. Anderson, April 5, 1945, Anderson to Dice, July 26, 1945, f. IHB—Corresp. and Admin. Material 1945, box 1, DHG. Reed, "Reactivation of the Dight Institute," 1; Resta, "Historical Perspective: Sheldon Reed and 50 years of genetic counseling," 375. On Reed, see: Stern, *Telling Genes*; Resta, "Historical Perspective"; Resta and Paul, "Historical Aspects of Medical Genetics."
31. Department of Human Genetics Collection, f. Inst of Hum Biol—Corresp and Administrative Material—1941–1942, box 1, Bentley Library, University of

Michigan. Muller was apparently being considered by Michigan as well. Neel, too, weighed the advantages of Michigan vs. Cold Spring Harbor and decided he needed the medical environment of the Heredity Clinic; Neel, *Physician to the Gene Pool*, 24–25.

32. "Gold mine in heredity": Allan to Paul David, November 10, 1941, f. 1941, box N307-1, Allan Papers. "Review of the Year's Activity, July 1941–July 1942," f. 1942–2 Jun–Dec, box N307-1, Allan papers. "Greatest natural experiment": Allan to Snyder, July 5, 1942, f. 1942-2 Jun–Dec, box N307-1, Allan Papers.
33. Allan to Osborn, December 1, 1941, f. 1941, box N307-1, Allan Papers.
34. Herndon to Allan, April 7, 1941, f. 1941, box N307-1, Allan Papers.
35. Herndon to Allan, July 25, 1941, f. 1941, box N307-1, Allan Papers.
36. Dudley to Herndon, May 14, 1942, f. 1942-1 (Jan.–May); Allan to Herndon, April 22, 1942, f. 1942-1 (Jan.–May); Allan to Snyder, December 12, 1942, f. 1942-2, box N307-1, Allan Papers.
37. Allan to Snyder, December 12, 1942, f. 1942-2; Allan to Paul David, November 10, 1941, f. 1941; Allan to Snyder, December 2, 1942, f. 1942-2 Jun–Dec, box N307-1, Allan Papers.
38. Herndon 16; Schoen, *Choice and Coercion*, 98–102.
39. Allan to Paul David, July 17 1942, f. 1942-2, box N307-1, Allan Papers.
40. "Report of Department of Medical Genetics, May 1, 1943 to May 1, 1944," f. Carnegie Corporation, box 7, Allan Papers; Schoen, *Choice and Coercion*, 30–43, chapter 2; Allan to Herndon, April 22, 1942, postscript April 25, f. 1942–1, box N307-1, Allan Papers.
41. Herndon to Florence Dudley, November 3, 1942; Herndon to Allan, September 22, 1942, f. 1942-2 Jun–Dec., box N307-1; "Report of Department of Medical Genetics May 1, 1943, to May 1, 1944," f. Carnegie Corporation, box 7, Allan Papers.
42. "Report of Department of Medical Genetics May 1, 1946, to May 1, 1947," f. Carnegie Corporation, box 7, Allan Papers.
43. Dice to Raymond W. Waggoner, Neuropsychiatric Institute, University of Michigan, April 27, 1945, f. IHB—Corresp. and Admin. Material 1945; NIH research grant application, f. Institute of Human Biology—Correspondence and administrative materials—1954, box 1, DHG; "Current Research Projects of LVB Staff, Aug. 1946," f. Laboratory of Vertebrate Biology Meeting Minutes, 1945–1952, box 5, Dice Papers.
44. Referrals: Untitled document, July 8, 1944, f. Institute of Human Biology—Correspondence and Administrative Material—1943–1944, box 1, DHG. Cotterman, "Status Bonnevie-Ullrich"; Cotterman, Falls, and Neel, "Some Hereditary Diseases"; Smyth, Cotterman, and Freyberg, "Genetics of Gout and Hyperuricaemia"; Smyth, Cotterman, and Freyberg, "Genetics of Gout and Hyperuricaemia: An Analysis"; Cotterman and Falls, "Unilateral Developmental Anomalies in Sisters"; Falls, Kruse, and Cotterman, "Three Cases of Marcus Gunn Phenomenon."

45. Reed, "Short History of Genetic Counseling"; Falls and Cotterman, "Genetic Studies on Ectopia Lentis," 11. On Reed, see Resta, "Historical Perspective"; Stern, *Telling Genes*. The small Harold Falls collection—just one box—at the Bentley Library, University of Michigan, contains a good set of Falls's reprints.
46. "Procedures Manual, 1946," f. Institute of Human Biology Procedures Manual, 1946, box 1, DHG.
47. Whitney, *Family Treasures*.
48. "Anthropometric measurements," "General neurological examination," and "Physical Examination, sheet E-1," f. Institute of Human Biology Procedures Manual, 1946, box 1, DHG.
49. For example, in 1945 Dice wrote to Harvard's Clark Heath, "It is our aim, as it seems to be yours, to find as many ways as possible for accurately measuring the human organism. We also are particularly interested in 'normal health and mentality,' though most of the cases we receive deal with physical or mental defects and illnesses" (Dice to Heath, October 9, 1945, f. IHB—Corresp. and Admin. Material 1945, box 1, DHG).
50. "Anthropometric measurements," "General neurological examination," and "Physical Examination, sheet E-1," f. Institute of Human Biology Procedures Manual, 1946, box 1, DHG.
51. www.mcgregorfund.org, last viewed September 11, 2011. Dice to Goethe, November 8, 1951, f. Goethe, C. M., box 2, Dice Papers.
52. Minutes of Staff Meeting, Inst. Human Biology, July 27, 1951, f. Corresp. and Admin. Material—1951, box 1, DHG; James P. Adams to Mr. Wickliffe Draper, April 25, 1950, Regents Exhibit; Wickliffe Draper to the President of the University of Michigan, April 18, 1950, Regents Exhibit, Bentley Historical Library, University of Michigan. Thanks to Alex Stern for this insight and detective work. She recounts these events in Stern, *Telling Genes*, chapter 2.
53. *Der Tagesspiegel*, August 19, 1951, f. Ka–Kk, box 3, Dice Papers. For a description of the AMS, see "Institute of Human Biology, 1951–52," f. Correspondence and Administrative Materials, 1951, box 1, DHG.
54. Paul, "Did Eugenics Rest on an Elementary Mistake?"

5. How the Geneticists Learned to Start Worrying and Love Mutation

1. F. A. E. Crew to H. J. Muller, August 4, 1939; N. I. Vavilov to Crew, July 26, 1939; Crew to Vavilov, undated; O. L. Mohr to Bailie Edward and Sir Thomas Beare, undated, f. 7th Intl. Congress of Genetics, box 1, Series III, H. J. Muller Papers, Lilly Library, Indiana University (hereafter Muller Mss.). On the history of the Congress, see Krementsov, *International Science between the World Wars*, esp. chapters 2 and 3; Doel, Hoffmann, and Krementsov, "National States and International Science," 59–66; Soyfer, "Tragic History."
2. Muller, "Social Biology and Population Improvement"; Gruenberg, "Men and Mice at Edinburgh."

3. Steinberg, "Much Ado about Me," 253; Associated Press, "Hit off Hebrides"; "List of the American Passengers"; Schull, "James Van Gundia Neel," 4. Note that in the *New York Times* list of survivors Price's first name is given as "Bunson." Cotterman to Dice, June 9, 1945, f. IHB—Corresp. and Admin. Material 1945, box 1, DHG; Price to Muller, September 5 [1939], f. Price, Bronson 1939–1945, box 27, Muller Mss.
4. For Bridges's and Sturtevant's side of the debate, see Kohler, *Lords of the Fly*; Sturtevant, *History of Genetics*. For Muller's side, see Schwartz, *In Pursuit of the Gene*; Carlson, *Genes, Radiation, and Society*.
5. Muller, "Mutation," 107, 112; Timofeeff-Ressovky, Zimmer, and Delbrück, "Über die Natur der Genmutation und der Genstruktur." See also Deichmann, "Brief Review," 6–9; Paul, "Eugenics and the Left," 575–78.
6. Muller, *Out of the Night*; Paul, "Eugenics and the Left," 578–82; Carlson, *Genes, Radiation, and Society*, 204–34.
7. Strandskov, "Discussion."
8. See, e.g., Wiener, "Genetic Theory of the Rh Blood Groups"; Wiener, "Nomenclature of the RH Blood Types"; Wiener, "Genetics and Nomenclature of the Rh-Hr Blood Types."
9. All quotations in this paragraph from Snyder to Muller, July 18, 1945, f. Snyder, Laurence, H., 1938–1945, box 28, Muller Mss.
10. Snyder telephone interview; Allan, "Relationship of Eugenics to Public Health," 74.
11. Muller, "Pilgrim Trust Lecture," 31.
12. Ibid., 32.
13. Muller, "Production of Mutations."
14. Beatty, "Genetics in the Atomic Age," 298. "Real gamble": Neel, *Physician to the Gene Pool*, 17.
15. Cotterman to Dice, June 9, 1945; Dice to Cotterman, June 14, 1945, f. IHB—Corresp. and Admin. Material 1945, box 1, DHG.
16. Beatty, "Genetics in the Atomic Age," 285; Lindee, *Suffering Made Real*, 37.
17. Lindee, *Suffering Made Real*, 70–76.
18. "Genetic Effects of the Atomic Bombs," 333.
19. H. H. Strandskov to H. J. Muller, f. 1947, May 21–31, box 3, series I, Muller Mss.
20. Kay, *Molecular Vision of Life*, 220; Dejong-Lambert and Krementsov, "On Labels and Issues"; Doel, Hoffmann, and Krementsov, "National States and International Science"; Krementsov, "'Second Front' in Soviet Genetics"; Krementsov, "From 'Beastly Philosophy' to Medical Genetics."
21. United States Office of Scientific Research and Development and Vannevar Bush, *Science, the Endless Frontier*; Kevles, "National Science Foundation."
22. Beaglehole et al., *Statement on Race*; Brattain, "Race, racism, and antiracism," esp. 1394–95; Dunn, *Race and Biology*; Dunn, "Meeting of Physical Anthropologists and Geneticists"; Gormley, "Scientific Discrimination and the Activist Scientist"; Collopy, "'Race' Relations"; Provine, "Geneticists and the Biology of Race

Crossing"; UNESCO, "Statement on Race." On Bentley Glass, see Wolfe, "Organization Man and the Archive."

23. Muller to Strandskov, February 11, 1948, f. American Society of Human Genetics 1947–1948, May, box 3, series V.
24. Strandskov to Muller, August 8, 1948, f. ASHG 1948, Jun–Aug, box 3, series V, Muller Mss. Hooton quotation from Hooton, "Human Heredity," 40.
25. Neel to Muller, Snyder, and Strandskov, September 21, 1948, f. American Society of Human Genetics, 1948, June–Aug., box 3, series V, Muller Mss.
26. Program of first meeting of Human Genetics Society of America, f. American Society of Human Genetics, box 1, series IV, James V. Neel Papers, American Philosophical Society Library, Philadelphia.
27. Neel, "Detection of the Genetic Carriers of Hereditary Disease." On eugenics and carrier detection, see Paul, "Is Human Genetics Disguised Eugenics?" 74.
28. Muller, "Our Load of Mutations," 142.
29. Ibid., 146–50.
30. Ibid., 150.
31. Ibid., 163.
32. Ibid., 165–66, 169.
33. Quotations from Wallace and Dobzhansky, *Radiation, Genes, and Man*, 190–91; Glass, "Genetics in the Service of Man." On the classical-balance controversy, see Beatty, "Weighing the Risks"; Paul, " 'Our Load of Mutations' Revisited."
34. Wallace and Dobzhansky, *Radiation, Genes, and Man*; Osborn, *Future of Human Heredity*; Dobzhansky, Foreword to *Future of Human Heredity*, vi. On Osborn's role in the AES, see Mehler, "History of the American Eugenics Society," esp. chapters 1, 7.
35. Muller to Neel, September 24, 1948, f. Am Soc Hum Genetics 1948, Sep–Dec, box 4, series V, Muller Mss.
36. Strandskov to Muller, August 20, 1948, H. L. Shapiro to Muller, August 20, 1948, f. American Society of Human Genetics, 1948, June–Aug.; Dice to Strandskov, cc Muller, December 9, 1948, f. Am Soc Hum Genetics 1948, Sep–Dec, box 4, series V, Muller Mss.
37. Cotterman to Snyder, Muller, and Strandskov, October 8, 1948; A. J. Carlson to Strandskov, October 21, 1948, f. Am Soc Hum Genetics 1948, Sep–Dec, box 4, series V, Muller Mss.
38. Muller to Cotterman, January 17, 1949, f. Am Soc Hum Genetics 1949, Jan–July, box 4, series V, Muller Mss.
39. Muller, "Progress and Prospects in Human Genetics," 2; Cotterman to Muller, September 24, 1949, f. Am Soc Hum Genetics 1949, Jan–July, box 4, series V, Muller Mss.; Weiss, "After the Fall"; Deichmann, "Hans Nachtsheim."
40. Muller, "Progress and Prospects in Human Genetics," 14.
41. Ibid., 16–17.
42. Carel Goldschmidt to Nolan DC Lewis, November 30, 1949, f. ASHG 1949, Jan–May, box 4, series V, Muller Mss.

43. Jane Schneidewind to Martha Tracy, February 7, 1951, f. Corresp. and Admin. Material–1951, box 1, DHG; Cotterman to Dice, June 29, 1951, f. Cotterman, Charles, box 1, Dice Papers.
44. Strandskov to ASHG Board of Directors, March 11, 1951, f. ASHG programs, membership, constitution, 1950–51, box 4, series V, Muller Mss.
45. Strandskov to ASHG members, May 12, 1951; Strandskov to Draper, May 9, 1951; Strandskov to Dice, May 17, 1951; Dice to Strandskov, May 21, 1951, Dice to Strandskov, May 29, 1951, f. ASHG, box 1, Dice Papers.
46. F. Draper, Wickliffe, Manson Meads Collection, Wake Forest Medical Archives. See also Begos, "Against Their Will"; Meads, Bowman Gray School of Medicine, and North Carolina Baptist Hospital, *Miracle on Hawthorne Hill*, 51.
47. C. M. Goethe to Eldon Gardner, September 8, 1959, f. Secretary 1957–61, box 1, ASHG Papers, APS.
48. Olden to Muller, August 15, 1947; Muller to Olden, August 27, 1947; Muller to Olden, December 10, 1947, f. Birthright 1947; also f. Birthright, Inc., 1948–49, box 4, series V, Muller Mss. For Birthright's campaign in the South during the fifties, see Schoen, "Reassessing Eugenic Sterilization," 148.
49. Neel, "Medicine's Genetic Horizons," 473–75.
50. Soraya de Chadarevian, "Mice and the Reactor," argues, from British evidence, for the importance of atomic links in the rise of postwar genetics.

6. Getting Their Organ

1. http://www.hopkinsmedicine.org/Press_releases/2008/07_23_08.html; "A Tribute to Victor A. McKusick, M.D., 'Father of Medical Genetics,'" http://www.marchofdimes.com/aboutus/22684_31024.asp; "NHGRI Pays Tribute to a Giant of Medical Genetics, Renowned Father of Medical Genetics, Victor Almon McKusick, 1921–2008," http://www.genome.gov/27527148.
2. McKusick, "Growth and Development of Human Genetics," 261; Penrose, "Future Possibilities in Human Genetics," 166–67.
3. Gaudillière, "Making Heredity in Mice and Men"; Lindee, "Genetic Disease in the 1960s."
4. Kevles, *In the Name of Eugenics*, chapter 16; Lindee, "Squashed Spiders"; McKusick, "Growth and Development of Human Genetics," 262.
5. Valle, "Victor Almon McKusick"; Vincent L. McKusick interview. Quotations from Victor A. McKusick interview.
6. Jeghers, McKusick, and Katz, "Generalized Intestinal Polyposis."
7. "I got training": Victor A. McKusick interview. Ford Walker's institute: N. F. Walker to Dice, June 29, 1951, f. WA–WD, box 4, Dice Papers. Dunn's institute: Dunn to Dice, August, 13, 1951, f. Cotterman, Charles, box 1, Dice Papers.
8. McKusick interview; McKusick, *Heritable Disorders of Connective Tissue*.
9. Kemp, Hauge, and Harvald, *Proceedings of the First International Congress of Human Genetics*. "Very defining experience": Victor A. McKusick interview. Alessandro Portelli writes of "condensation" of time in oral narratives; Portelli, *Death of Luigi*

Trastulli, 13–16. I am using a similar idea here, but with a different metaphor to connote both the bidirectionality of time shifting and the tendency to create "watershed moments."

10. Bomholt, "Address at the Opening of the First International Congress of Human Genetics"; Kemp, "Address at the Opening of the First International Congress of Human Genetics."
11. Kemp, in Kemp, Hauge, and Harvald, *Proceedings of the First International Congress of Human Genetics*, 456. Note: I cite titled articles from this volume as such; quotations from transcribed panel discussions are cited with the volume and page numbers only.
12. Ibid., 457. One interview source, speaking off the record, expressed his frustration at being unable to implement such a registry.
13. Lyon, "Gene Action in the X-Chromosome"; Fuchs and Riis, "Antenatal Sex Determination"; Barr and Bertram, "Morphological Distinction."
14. Riis et al., "Cytological Sex Determination in disorders"; Fuchs et al., "Antenatal Detection of Hereditary Diseases," 261, 262.
15. Cowan, "Women's Roles," 37; Riis and Fuchs, "Antenatal Determination of Fetal Sex."
16. Tjio and Levan, "Chromosome Number of Man"; Victor A. McKusick interview; Ford and Hamerton, "Chromosomes of Man."
17. Kottler, "From 48 to 46"; Martin, "Can't Any Body Count?"; Harper, *First Years of Human Chromosomes*, chapter 2. On human cytogenetics generally, see Lindee, *Moments of Truth in Genetic Medicine*, 98–101; Harper, *First Years of Human Chromosomes*; Harper, *Short History of Medical Genetics*. On HeLa, see Skloot, *Immortal Life of Henrietta Lacks*; Landecker, *Culturing Life*. On deionized water, see Hsu, "Mammalian Chromosomes in Vitro"; Hsu, *Human and Mammalian Cytogenetics*. The squash technique was developed by Belling, "Counting Chromosomes in Pollen-Mother Cells."
18. Soraya, "Mice and the Reactor"; Jacobs et al., "Existence of the Human 'Super Female' "; Jacobs et al., "Chromosomal Sex"; Ford et al., "Sex-Chromosome Anomaly"; Jacobs and Strong, "Case of Human Intersexuality"; Maclean et al., "Survey of Sex-Chromosome Abnormalities"; Court Brown, "Sex Chromosomes and the Law."
19. Nowell quotation: Nowell interview. For examples of ambiguities in chromosome assignment, see Ford et al., "Chromosomes Showing Both Mongolism and the Klinefelter Syndrome"; Jacobs et al., "Somatic Chromosomes in mongolism"; Lejeune, Gautier, and Turpin, "Etude des chromosomes somatiques"; Lejeune, Turpin, and Gautier, "Le mongolisme"; Patau et al., "Multiple Congenital Anomaly"; Edwards et al., "New Trisomic Syndrome." On the Denver conference, see "Proposed Standard System," 320; Lindee, *Moments of Truth in Genetic Medicine*, 105–8. Trisomy D and trisomy E: McKusick, "Lumpers and Splitters," 300. Harper, *First Years of Human Chromosomes*, chapter 3 traces the discovery of trisomy 21.

20. Guthrie, "Blood Screening for Phenylketonuria"; Guthrie and Susi, "Simple Phenylalanine Method"; Culley et al., "Paper Chromatographic Estimation"; Clow, Scriver, and Davies, "Results of Mass Screening for Hyperaminoacidemias"; Paul, "History of Newborn Screening"; Lindee, "Babies' Blood," 34–37.
21. Williams, *Biochemical Individuality*; Williams, *Why Human Genetics?* 2: 173. Lindee, "Genetic Disease in the 1960s," discusses this idea but places it later, in the 1960s.
22. Garrod, "Incidence of Alkaptonuria."
23. Childs interview.
24. Fraser interview.
25. For examples, see interviews with Charles Scriver, Haig Kazazian, David Valle, and Barbara Migeon, in the Oral History of Human Genetics Collection (http://ohhgp.pendari.com).
26. Childs, *Genetic Medicine.*
27. Edelson, "Adopting Osler's Principles."
28. Harvey, McKusick, and Stobo, *Osler's Legacy*, 75.
29. Victor A. McKusick interview.
30. "McKusick's followup study," box 433591630, Constitution Clinic Miscellaneous Recs, Raymond Pearl Collection, Chesney Archives, Johns Hopkins University. On McKusick's studies on the Amish, see Lindee, "Provenance and the Pedigree."
31. McKusick, "Persisting Memories of Cyril Clarke in Baltimore"; Kohler, *Lords of the Fly*, chapter 2; Lindee, "Squashed Spiders"; Kevles, *In the Name of Eugenics*, chapter 16.
32. Migeon interview; Ferguson-Smith, "Techniques of Human Cytogenetics." For comparison with maize, see Comfort, *Tangled Field*, esp. chapter 3; for *Drosophila*, see Kohler, *Lords of the Fly*, chapter 3.
33. See, for example, Ferguson-Smith and Johnston, "Chromosome Abnormalities"; Ferguson-Smith and Johnston, "Human Chromosomes"; Alexander and Ferguson-Smith, "Chromosomal Studies"; Johnston et al., "Triple-X Syndrome"; Ferguson-Smith, "Sex Chromatin Anomalies"; Ferguson-Smith et al., "Abnormal Metacentric Chromosome."
34. Nowell, Hungerford, and Brooks, "Chromosomal Characteristics"; Baikie et al., "Chromosome Studies in Human Leukaemia"; Baikie et al., "Possible Specific Chromosome Abnormality"; Nowell and Hungerford, "Minute Chromosome"; Nowell and Hungerford, "Aetiology of Leukaemia"; Nowell and Hungerford, "Chromosome Studies"; Nowell interview.
35. Nowell interview; Sturtevant, "Genetic Factors."
36. Motulsky interview.
37. Ibid.
38. Ibid.
39. Motulsky, "Drug Reactions," 836–37; Vogel, "Moderne Problem." On the early history of primaquine and other synthetic antimalarials, see Slater, "Malaria Chemotherapy"; Comfort, "Prisoner as Model Organism."

40. Brink, "Early History of Genetics."
41. Brink interview, 45–46; Brink, "Early History of Genetics"; Lederberg interview.
42. Lederberg, "Some Early Stirrings of Concern," 4.
43. Oliver, *Making the Modern Medical School*; Lederberg, "Memorandum to Dean Bowers."
44. Lederberg interview.
45. Bowers, Lederberg, and Mortenson, "Proposal for a Program."
46. McKusick, "Mendelian Inheritance in Man."
47. Gaudillière, "Making Heredity in Mice and Men."
48. McKusick to Abbey, Boyer, Childs, Lilienfeld, Sidbury, and Young, September 29, 1959, f. Roscoe B. Jackson Memorial Laboratory, ó 1960 (inclusive), box 110B17–18, Victor McKusick Papers, Alan Mason Chesney Archives, Johns Hopkins University, Baltimore (hereafter VAM).
49. "One of the leading functions": McKusick to John L. Fuller, November 14, 1960, f Bar Harbor 1961, ó 1960–1961 (inclusive), box 109C15; "Some criticize": McKusick to Fuller, undated, f. Roscoe B. Jackson Memorial Laboratory, ó 1960 (inclusive), box 110B17–18; "What I have in mind": Fuller to McKusick, November 8, 1960 [handwritten notes at bottom], f. Bar Harbor 1961, 1960–1961 (inclusive), box 109C15, VAM.
50. "Sole support": Victor A. McKusick interview. Eugenics Society presence: f. Correspondence re Bar Harbor (Miscellaneous) 1960–1961 (inclusive), box 109C15, VAM.
51. Evaluation questionnaires, box 110B17–18, VAM.
52. Margery Shaw to McKusick, August 24, 1960, f. Bar Harbor 1961, ó 1960–1961 (inclusive), box 109C15; McKusick to David M. Bonner, March 27, 1962, General Correspondence—Bar Harbor, ó 1962–1965 (inclusive), box 109–10; Park S. Gerald to McKusick, October 16, 1961, f. Correspondence re Bar Harbor (Miscellaneous), fox 109C15, VAM.
53. F. Bar Harbor 1961, 1960–1961 (inclusive), VAM.
54. Richard C. Lewontin, "Summation, Bar Harbor Short Course on Human Genetics, 1968," box 109–10, folder 431969615, VAM.
55. F. Bar Harbor 1968 correspondence, ó 1967–1968, box 109–10, VAM.
56. McKusick, "Birth Defects," 407.

7. Genetics without Sex

1. Watson and Stent, *Double Helix*, 130.
2. Landecker, *Culturing Life*, 34, chapters 3, 4; Weller and Robbins, "John Franklin Enders"; Skloot, *Immortal Life of Henrietta Lacks*. In my discussion of model organisms, I am drawing on a rich literature, including Lederman et al., "Right Organism for the Job"; Summers, "How Bacteriophage Came to be Used"; Kohler, *Lords of the Fly*; Chadarevian, "Worms and Programmes"; Ankeny, "Fashioning Descriptive Models in Biology"; Todes, *Pavlov's Physiology Factory*; Creager, *Life of a Virus*; Rader, *Making Mice*; Comfort, "Prisoner as Model Organism."

3. Lederberg and Tatum, "Gene Recombination in Escherichia coli"; Zinder and Lederberg, "Genetic Exchange in Salmonella."
4. Haldane, "Some Alternatives to Sex," 21.
5. Did not take the world by storm: Olby, "Quiet Debut for the Double Helix"; Chadarevian, *Designs for Life*, 241–45; Chadarevian, "Mice and the Reactor." Cf. Gingras, "Revisiting the 'Quiet Debut.'" On Watson's posthelical postdoc, see Creager and Morgan, "After the Double Helix." On Meselson and Stahl, see Holmes, *Meselson, Stahl, and the Replication of DNA*. On the genetic code, see Judson, *Eighth Day of Creation*, 455–69; Hayes, "Invention of the Genetic Code"; Matthaei and Nirenberg, "Characteristics and Stabilization"; Nirenberg and Matthaei, "Dependence of Cell-Free Protein Synthesis." On Jacob and Monod, see Mueller-Hill, *Lac Operon*; Jacob et al., "L'opéron." On Rosalind Franklin, see Maddox, *Rosalind Franklin*.
6. McElheny, *Watson and DNA*, 101–8.
7. Sayre, *Rosalind Franklin and DNA*; Maddox, *Rosalind Franklin*.
8. Stent, *Coming of the Golden Age*.
9. On Penrose's biography, see Harris, "Lionel Sharples Penrose"; Berg, "Lionel Sharples Penrose"; Laxova, "Lionel Sharples Penrose"; Watt, "L. S. Penrose"; Watt, "Lionel Penrose, F.R.S."; Smith, *Lionel Sharples Penrose*. On Colchester, see Kevles, *In the Name of Eugenics*, 148–63.
10. Penrose, *Clinical and Genetic Study*, 70.
11. F. Penrose, LS, box 3, Lee R. Dice Papers, Bentley Library, University of Michigan; Penrose, "Phenylketonuria."
12. Penrose, "Phenylketonuria."
13. Fraser interview; Motulsky interview; Hirschhorn interview; Scriver interview.
14. Harris, "The 'Inborn Errors' Today," 180.
15. Harris, "Enzyme Polymorphisms in Man," 308.
16. Scriver interview; Dent, "Paper Chromatography and Medicine," 238. "Most horrid smell": http://links.math.rpi.edu/devmodules/graph_isomorphism/html/Naming.html; Scriver interview. For an explanation of the method, see Dent, "Paper Chromatography and Medicine."
17. Kay, "Laboratory Technology and Biological Knowledge"; Chiang, "Laboratory Technology," 502–5; Neel, "Inheritance of Sickle Cell Anemia"; Pauling et al., "Sickle Cell Anemia"; Ingram, "Gene Mutations in Human Haemoglobin"; Strasser, "Linus Pauling's 'Molecular Diseases'"; Feldman and Tauber, "Sickle Cell Anemia."
18. Scriver interview.
19. Kunkel and Slater, "Zone Electrophoresis"; Smithies interview; Consden, Gordon, and Martin, "Qualitative Analysis of Proteins"; Smithies, "Zone Electrophoresis in Starch Gels"; Smithies, "Improved Procedure"; Chiang, "Laboratory Technology," 510–11.
20. Smithies interview. See also Smithies, "Starch-Gel Electrophoresis."
21. Chiang, "Laboratory Technology," 511–13.

22. On the history of the information model of biology, see Kay, *Who Wrote the Book of Life?*
23. Burian and Gayon, "French School of Genetics"; Judson, *Eighth Day of Creation*, 272, 390–408; Osmundsen, "Biologists Hopeful of Solving Secrets of Heredity."
24. Weisberger, "Induction of Altered Globin Synthesis"; Szybalska and Szybalski, "Genetics of Human Cell Line"; Szybalski, Szybalska, and Ragni, "Genetic Studies with Human Cell Lines"; McKusick, "Medical Genetics 1962," 458.
25. "Dictates . . . controls"; "juggling the code itself": "A.M.A. Cites Gains by Medicine in '62"; "Newton . . . Beethoven . . . Einstein": Laurence, "Nobel Winners."
26. Lederberg, "Molecular Biology, Eugenics, and Euphenics," 428; Lederberg, "Experimental Genetics and Human Evolution," 521.
27. Watkins, *On the Pill*, chapter 1; Paul, "History of Newborn Screening," 141.
28. "Topics in human genetics with social implications" (undated, but prob. c. 1970), f. ASHG Ethics, box 2, Neel papers, APS.
29. Ptashne, "Isolation of the lambda Phage Repressor"; Reinhold, "Scientists Isolate a Gene."
30. Watson, "Moving toward the Clonal Man."
31. For an example drawn from arguments against genetic enhancement, see Resnik and Vorhaus, "Genetic Modification and Genetic Determinism."
32. Glass and Comfort, Building Arcadia, chapter 11, p. 27; but see McElheny, *Watson and DNA*, 162, for a slightly different story.
33. Yi, "Cancer, Viruses, and Mass Migration"; Wright, "Recombinant DNA Technology"; Wade, "Special Virus Cancer Program."
34. Schmidt, "Five Years into the National Cancer Program," 238; Office of Technology Assessment, *Technology Transfer*, 83; Rettig, *Cancer Crusade*, Afterword.
35. Krimsky, *Genetic Alchemy*, 29.
36. Berg et al., "Potential Biohazards of Recombinant DNA Molecules."
37. Watson and Tooze, *DNA Story*, 91–94. See also Swazey, Sorenson, and Wong, "Risks and Benefits, Rights and Responsibilities." On how Cambridge became the obvious place for such a debate, see Emrich, "Dr. Genelove."
38. See, e.g., Berg, "Reflections on Asilomar 2 at Asilomar 3"; Watson, "In Further Defense of DNA," 58. Many relevant documents are conveniently collected and reproduced in facsimile by Watson himself and John Tooze, in *DNA Story*.
39. Annas and Coyne, "'Fitness' for Birth and Reproduction"; Jacobs et al., "Aggressive Behavior"; Court Brown, "Sex Chromosomes and the Law"; Telfer et al., "Incidence of Gross Chromosomal Errors"; Lederberg, "Chromosomes and Crime."
40. Walzer, Breau, and Gerald, "Chromosome Survey"; Walzer interview.
41. Beckwith and King, "XYY Syndrome"; Beckwith et al., "Harvard XYY Study"; Beckwith and Miller, "XYY Male"; Culliton, "Patients' Rights"; Culliton, "XYY"; Steinfels and Levine, "XYY Controversy"; Roblin, "Boston XYY Case"; Beckwith interview.
42. Barski, Sorieul, and Cornefert, "Production dans des cultures in vitro." See also Harris, *Cells of the Body*, 121–22; Zallen and Burian, "Beginnings of Somatic Cell Hybridization."

43. Weiss and Green, "Human-Mouse Hybrid Cell Lines"; Zallen and Burian, "Beginnings of Somatic Cell Hybridization."
44. Ruddle interview; Weiss and Green, "Human-Mouse Hybrid Cell Lines."
45. Ruddle interview; McKusick and Ruddle, "Status of the Gene Map," 402; McKusick, "Human Gene Map," 364; Solomon and Bodmer, Introduction to *Cytogenetics and Cellular Genetics* 58.
46. Kuska, "Beer, Bethesda, and Biology"; Victor A. McKusick interview; Ruddle interview.
47. Kan and Dozy, "Polymorphism of DNA Sequence."
48. Botstein et al., "Construction of a Genetic Linkage Map."
49. Davies et al., "Cloning of a Representative Genomic Library" Murray et al., "Linkage Relationship of a Cloned DNA Sequence"; Harper et al., "Use of Linked DNA Polymorphisms"; Koenig et al., "Complete Cloning of the Duchenne Muscular Dystrophy (DMD) cDNA."
50. Drumm et al., "Physical Mapping of the Cystic Fibrosis Region"; Kerem et al., "Identification of the Cystic Fibrosis Gene"; Riordan et al., "Identification of the Cystic Fibrosis Gene"; Rommens et al., "Identification of the Cystic Fibrosis Gene."
51. Wexler, *Mapping Fate.*
52. Davies and White, *Breakthrough.*
53. Schmeck, "Virus Is Injected into 2 Children"; Friedmann and Roblin, "Gene Therapy for Human Genetic Disease?"; Lyon and Gorner, *Altered Fates,* 68–76, quotation from 74.
54. Stern, *Telling Genes,* esp. chap. 3.
55. Lyon and Gorner, *Altered Fates,* 21.
56. Sinsheimer, quoted in Anderson, "Genetics and Human Malleability," 4.
57. Ibid., 24.
58. Guterman, "Gene-Therapy Pioneer Convicted of Abuse."
59. Stolberg, "Biotech Death of Jesse Gelsinger," 49–50; Weiss and Nelson, "Teen Dies Undergoing Experimental Gene Therapy"; Couzin-Frankel, "Genetics"; Kaiser, "Gene Therapy"; Naldini, "Medicine."
60. Keller, *Century of the Gene,* 6–8.
61. Nelkin and Lindee, *DNA Mystique,* chapters 4 and 5; "true self," 42, "power of heredity," 97; Roof, *Poetics of DNA,* chapter 3.
62. Wade, "Genome of DNA Discoverer Is Deciphered." Watson's DNA can be viewed at http://jimwatsonsequence.cshl.edu.

Epilogue

1. Ridley, Foreword to *Davenport's Dream,* ix; Carlson, "Eugenic World of Charles Davenport," 76; Olson, "Davenport's Dream," 78.
2. Anderson, "Human Gene Therapy," 89–90; Anderson, "Human Gene Therapy."
3. Mehlman, *Price of Perfection,* chapter 1.

4. Lander and Schork, "Genetic Dissection of Complex Traits," 2039; Jordan, "Place of Eugenics," 396; Hood, "Biology and Medicine in the Twenty-First Century," 158.
5. Notes from Johns Hopkins Medical School Council (Faculty Senate) meetings, 2009–11. For the drug side of this story, see Greene, *Prescribing by Numbers*.
6. Ridley, "Davenport's Dream"; Agar, "Liberal Eugenics"; Agar, *Liberal Eugenics*.
7. Harris, *Enhancing Evolution*, 4–5.
8. Galton, "Hereditary Improvement," 120; Galton, "Possible Improvement of the Human Breed" (1901).

BIBLIOGRAPHY

Note: For articles in some major journals, the Digital Object Identifier (DOI) is given. The DOI system is an ISO International Standard that assigns a unique, permanent name to electronic digital property. Even if the location of the object changes, the DOI remains consistent. Articles can be retrieved by their DOI via any search engine. For more information see http://doi.org.

Ackerknecht, Erwin. "Diathesis: The Word and the Concept in Medical History." *Bulletin of the History of Medicine* 56 (1982): 317–25.

Agar, Nicholas. "Liberal Eugenics." *Public Affairs Quarterly* 12, no. 2 (1998): 137–55.

———. *Liberal Eugenics: In Defence of Human Enhancement*. Malden, MA: Blackwell, 2005.

Alexander, D. S., and M. A. Ferguson-Smith. "Chromosomal Studies in Some Variants of Male Pseudohermaphroditism." *Pediatrics* 28 (1961): 758–63.

Allan, William. "The Effect of Emetine on Entamoeba Histolytica in Stools." *Journal of Pharmacology and Experimental Therapeutics* 16, no. 1 (1921): 21–33.

———. "Eugenic Significance of Retinitis Pigmentosa." *Archives of Ophthalmology* 18 (1937): 938–47.

———. "Medicine's Need of Eugenics." *Southern Medical and Surgical Journal* 98 (1936): 416–17.

———. "Preventing Hereditary Diseases That Wreck Childhood." *Journal of the Association of American Medical Colleges* 14 (1939): 231–34.

———. "The Relationship of Eugenics to Public Health." *Eugenical News* 21 (1936): 73–75.

Allen, Garland E. "The Eugenics Record Office at Cold Spring Harbor, 1910–1940." *Osiris* 2 (1986): 225–64.

———. "Old Wine in New Bottles: From Eugenics to Population Control in the Work of Raymond Pearl." In *Expansion of American Biology*, ed. Keith Rodney Benson,

Jane Maienschein, and Ronald Rainger, 231–61. New Brunswick, NJ: Rutgers University Press, 1991.

Allen, Robert Loring. *Irving Fisher: A Biography.* Cambridge, MA: Blackwell, 1993.

"A.M.A. Cites Gains by Medicine in '62: But Breakthroughs Require Further Study, It Says Defects Explained." *New York Times,* January 2, 1963.

Anbinder, Tyler. *Five Points: The 19th-Century New York City Neighborhood That Invented Tap Dance, Stole Elections, and Became the World's Most Notorious Slum.* New York: Free Press, 2001.

Anderson, W. F. "Genetics and Human Malleability." *Hastings Center Report* 20, no. 1 (1990): 21–24.

———. "Human Gene Therapy: Scientific and Ethical Considerations." *Journal of Medicine and Philosophy* 10, no. 3 (1985): 275–91.

———. "Human Gene Therapy: Why Draw a Line?" *Journal of Medicine and Philosophy* 14, no. 6 (1989): 681–93.

Ankeny, Rachel A. "Fashioning Descriptive Models in Biology: Of Worms and Wiring Diagrams." *Philosophy of Science* 67 (2000): suppl. S260–72.

Annas, G. J., and B. Coyne. " 'Fitness' for Birth and Reproduction: Legal Implications of Genetic Screening." *Family Law Quarterly* 9, no. 3 (1975): 463–89.

[Anonymous]. "Herndon." *Winston-Salem Journal,* March 31, 1988.

Aronson, J. K. "Francis Galton and the Invention of Terms for Quantiles." *Journal of Clinical Epidemiology* 54, no. 12 (2001): 1191–94.

Associated Press. "Hit Off Hebrides; Ship Bound for Canada Carried Some Children among Americans; Capital Is Shocked; President's Aide Notes Liner Had Refugees, Not Munitions; 292 Americans Were Aboard; White House Is Informed; Liner Torpedoed; 1,400 Passengers; Montreal Not Informed; Vessel Built in 1923." *New York Times,* September 4, 1939, 1–2.

Baikie, A. G., W. M. Court-Brown, K. E. Buckton, D. G. Harnden, P. A. Jacobs, and I. M. Tough. "A Possible Specific Chromosome Abnormality in Human Chronic Myeloid Leukaemia." *Science* 188 (1960): 1165–66. doi: 10.1038/1881165a0.

Baikie, A. G., W. M. Court Brown, P. A. Jacobs, and J. S. Milne. "Chromosome Studies in Human Leukaemia." *Lancet* 274, no. 7100 (1959): 425–28.

Barker, Lewellys F. Foreword to *Eugenics: Twelve University Lectures,* ed. Lucy James Wilson, ix–xiii. New York: Dodd, Mead, 1914.

———. "Heredity in the Clinic." *American Journal of the Medical Sciences* 173, no. 5 (1927): 597–605.

Barr, Murray L., and E. G. Bertram. "A Morphological Distinction between Neurones of the Male and Female, and the Behaviour of the Nucleolar Satellite during Accelerated Nucleoprotein Synthesis." *Nature* 163 (1949): 676–77. doi: 10.1038/163676a0.

Barski, G., S. Sorieul, and F. Cornefert. "Production dans des cultures in vitro de deux souches cellulaires en association, de cellules de caractere 'hybride'." *Comptes Rendus de l'Académie des Sciences* 251 (1960): 1825–27.

Bateson, William. *Mendel's Principles of Heredity: A Defence.* Cambridge: Cambridge University Press, 1902.

———. *William Bateson, F. R. S., Naturalist: His Essays and Addresses, Together with a Short Account of His Life*. Cambridge: Cambridge University Press, 1928.

Bateson, William, and Reginald Crundall Punnett. "On the Inter-Relations of Genetic Factors." *Proceedings of the Royal Society of London*. Series B 84 (1911): 3–8.

Bateson, William, and E. R. Saunders. "The Facts of Heredity in the Light of Mendel's Discovery." *Reports to the Evolution Committee of the Royal Society* 1 (1902): 125–60.

Beaglehole, Ernest, Juan Comas, L. A. Costa Pinto, Franklin Frazier, Morris Ginsberg, Humayun Kabir, Claude Levi-Strauss, and Ashley Montagu. *Statement on Race* (UNESCO). Lake Success, NY: United Nations Department of Public Information, Press, and Publications Bureau, 1950.

Bearn, A[lexander]. G. "Archibald Edward Garrod, the Reluctant Geneticist." *Genetics* 137, no. 1 (1994): 1–4.

———. *Archibald Garrod and the Individuality of Man*. Oxford: Clarendon, 1993.

Bearn, A. G., and E. D. Miller. "Archibald Garrod and the Development of the Concept of Inborn Errors of Metabolism." *Bulletin of the History of Medicine* 53, no. 3 (1979): 315–28.

Beatty, John. "Genetics in the Atomic Age: The Atomic Bomb Casualty Commission, 1947–1956." In *Expansion of American Biology*, ed. Keith Rodney Benson, Jane Maienschein, and Ronald Rainger, 284–324. New Brunswick, NJ: Rutgers University Press, 1991.

———. "Weighing the Risks: Stalemate in the Classical/Balance Controversy." *Journal of the History of Biology* 20 (1987): 289–319.

Beckwith, Jonathan R. Interview with Ami Karlage, August 10, 2008, Cambridge, MA. Oral History of Human Genetics Project, UCLA and Johns Hopkins University, http://ohhgp.pendari.com.

Beckwith, J[onathan]., et al. "Harvard XYY Study." *Science* 187, (1975): 298–99. doi: 10.1126/science.11643259.

Beckwith, Jonathan R., and J. King. "The XYY Syndrome: A Dangerous Myth." *New Scientist* 64, no. 923 (1974): 474–76.

Beckwith, J., and L. Miller. "The XYY Male: The Making of a Myth." *Harvard Magazine* 79, no. 2 (1976): 30–33.

Begos, Kevin. "Against Their Will: North Carolina's Sterilization Program." *Winston-Salem Journal*, December 8–12, 2002, http://againsttheirwill.journalnow.com/.

Belling, John. "On Counting Chromosomes in Pollen-Mother Cells." *American Naturalist* 55 (1921): 573–74.

Benson, Keith R. "The Naples Stazione Zoologica and Its Impact on the Emergence of American Marine Biology." *Journal of the History of Biology* 21 (1988): 331–41.

Berg, Joseph M. "Lionel Sharples Penrose (1898–1972): Aspects of the Man and His Works, with Particular Reference to His Undertakings in the Fields of Intellectual Disability and Mental Disorder." *Journal of Intellectual Disability Research* 42 (1998): 104–11.

Berg, Paul. "Reflections on Asilomar 2 at Asilomar 3: Twenty-Five Years Later." *Perspectives in Biology and Medicine* 44, no. 2 (2001): 183–85.

Berg, Paul, David Baltimore, Herbert W. Boyer, Stanley N. Cohen, Ronald W. Davis, David S. Hogness, Daniel Nathans, et al. "Potential Biohazards of Recombinant DNA Molecules." *Science* 185 (1974): 303. doi: 10.1126/science.185.4148.303.

Bernard, Claude. *An Introduction to the Study of Experimental Medicine.* 1865; New York: Dover, 1957.

Bix, A. S. "Experiences and Voices of Eugenics Field-Workers: 'Women's Work' in Biology." *Social Studies of Science* 27, no. 4 (1997): 625–68.

Blakeslee, Albert F. "Genetics of Sensory Thresholds: Taste for Phenylthiocarbamide." *Proceedings of the National Academy of Sciences U S A* 18 (1932): 120–26.

"Blind to a Nightmare." *Chicago Sun-Times,* August 31, 2003, 21.

Blodgett, Jan. "North Carolina Medical College." Davidson College Archives, http://forum.davidson.edu/archives/encyclopedia/north-carolina-medical-college/.

Bomholt, Julius. "Address at the Opening of the First International Congress of Human Genetics, Aug. 1st, 1956." In Kemp, Hauge, and Harvald, *Proceedings of the First International Congress of Human Genetics,* xi–xii.

Botstein, D., R. L. White, M. Skolnick, and R. W. Davis. "Construction of a Genetic Linkage Map in Man Using Restriction Fragment Length Polymorphisms." *American Journal of Human Genetics* 32, no. 3 (1980): 314–31.

Bowers, John Z., Joshua Lederberg, and O. A. Mortenson. "A Proposal for a Program in Human Genetics." Joshua Lederberg Papers, National Library of Medicine, Bethesda, MD, 1956.

Brattain, Michelle. "Race, Racism, and Antiracism: UNESCO and the Politics of Presenting Science to the Postwar Public." *American Historical Review* 107, no. 4 (2002): 1386–1413.

Brian, Thomas. *The Pisse-Prophet; or, Certaine Pisse-Pot Lectures: Wherein Are Newly Discovered the Old Fallacies, Deceit, and Jugling of the Pisse-Pot Science, Used by All Those (Whether Quacks and Empiricks, or Other Methodicall Physicians) Who Pretend Knowledge of Diseases, by the Urine, in Giving Judgement of the Same.* London: R. Thrale, 1637.

Brink, R. A. "Early History of Genetics at the University of Wisconsin–Madison." Joshua Lederberg Papers, National Library of Medicine, Bethesda, MD, 1974.

———. Interview with Donna S. Taylor, 1982, Madison, WI. University Archives Oral History Project, Steenbock Library, University of Wisconsin, Madison.

Broberg, Gunnar, and Nils Roll-Hansen. *Eugenics and the Welfare State: Sterilization Policy in Denmark, Sweden, Norway, and Finland.* East Lansing: Michigan State University Press, 2005.

Brookes, Martin. *Extreme Measures: The Dark Visions and Bright Ideas of Francis Galton.* New York: Bloomsbury, 2004.

Browne, E. Janet. *Charles Darwin: The Power of Place.* New York: Knopf, 2002.

Bulmer, M. G. *Francis Galton: Pioneer of Heredity and Biometry.* Baltimore: Johns Hopkins University Press, 2003.

Burbridge, D. "Francis Galton on Twins, Heredity, and Social Class." *British Journal of the History of Science* 34, no. 3 (2001): 323–40.

Burgio, G. R. "Diathesis and Predisposition: The Evolution of a Concept." *European Journal of Pediatrics* 155, no. 3 (1996): 163–64.

Burian, R. M., and J. Gayon. "The French School of Genetics: From Physiological and Population Genetics to Regulatory Molecular Genetics." *Annual Review of Genetics* 33 (1999): 313–49. doi: 10.1146/annurev.genet.33.1.313.

Burnham, John. *How Superstition Won and Science Lost: Popularizing Science and Health in the United States.* New Brunswick, NJ: Rutgers University Press, 1987.

Cannon, Walter B. *The Wisdom of the Body.* New York: Norton, 1932.

Carlson, Elof A[xel]. "The Eugenic World of Charles Davenport." In Witkowski, Inglis, and Davenport, *Davenport's Dream*, 59–76.

———. *Genes, Radiation, and Society: The Life and Work of H. J. Muller.* Ithaca, NY: Cornell University Press, 1981.

———. *The Unfit: A History of a Bad Idea.* Cold Spring Harbor, NY: Cold Spring Harbor Laboratory Press, 2001.

Castle, William E. "The Beginnings of Mendelism in America." In *Genetics in the Twentieth Century*, ed. L. C. Dunn, 59–76. New York: Macmillan, 1951.

Chadarevian, Soraya de. *Designs for Life: Molecular Biology after World War II.* New York: Cambridge University Press, 2002.

———. "Mice and the Reactor: The 'Genetics Experiment' in 1950s Britain." *Journal of the History of Biology* 39, no. 4 (2006): 707–35.

———. "Of Worms and Programmes: Caenorhabditis elegans and the Study of Development." *Studies in History and Philosophy of Biological and Biomedical Science* 29, no. 1 (1998): 81–105.

Chiang, H. H. "The Laboratory Technology of Discrete Molecular Separation: The Historical Development of Gel Electrophoresis and the Material Epistemology of Biomolecular Science, 1945–1970." *Journal of the History of Biology* 42, no. 3 (2009): 495–527.

Childs, Barton. *Genetic Medicine: A Logic of Disease.* Baltimore: Johns Hopkins University Press, 1999.

———. Interview with Andrea Maestrejuan, December 12, 2001, Baltimore. Oral History of Human Genetics Project, UCLA and Johns Hopkins University, http://ohhgp.pendari.com.

Clow, Carol L., Charles R. Scriver, and E. Davies. "Results of Mass Screening for Hyperaminoacidemias in the Newborn Infant." *American Journal of Diseases of Children* 117, no. 1 (1969): 48–53.

Cock, A. G. "Bateson's Two Toronto Addresses, 1921: 1. Chromosomal Skepticism. 2. Evolutionary Faith." *Journal of Heredity* 80 (1989): 91–99.

———. "William Bateson's Rejection and Eventual Acceptance of Chromosome Theory." *Annals of Science* 40 (1983): 19–59.

Coleman, William. "Bateson and Chromosomes: Conservative Thought in Science." *Centaurus* 15 (1970): 228–314.

Collopy, Peter. "'Race' Relations: Montagu, Dobzhansky, Coon, and the Divergence of Race Concepts." Joint Atlantic Seminar in the History of Biology (JAS-Bio), Drew University, 2009.

Comfort, Nathaniel. "The Prisoner as Model Organism: Malaria Research at Stateville Penitentiary." *Studies in History and Philosophy of Science, Part C: Studies in History and Philosophy of Biological and Biomedical Sciences* 40 (2009): 190–203.

———. *The Tangled Field: Barbara McClintock's Search for the Patterns of Genetic Control.* Cambridge: Harvard University Press, 2001.

———. "When Your Sources Talk Back: Toward a Multimodal Approach to Scientific Biography." *Journal of the History of Biology* 44: 651–59. doi: 10.1007/s10739-011-9273-9.

Consden, R., A. H. Gordon, and A. J. Martin. "Qualitative Analysis of Proteins: A Partition Chromatographic Method Using Paper." *Biochemical Journal* 38, no. 3 (1944): 224–32.

Cooke, Kathy. "False Gods and Red-Blooded Women: Reproduction, Environment, and American Character in the Early Twentieth Century." (Submitted.)

———. "The Limits of Heredity: Nature and Nurture in American Eugenics before 1915." *Journal of the History of Biology* 31, no. 2 (1998): 263–78.

Correns, Carl. "G. Mendel's Regel über das Verhalten der Nachkommenschaft der Rassenbastarde" (G. Mendel's law concerning the behavior of progeny of varietal hybrids). *Berichte der Deutschen Botanischen Gesellschaft* 18, no. 4 (1900): 158–68.

Cotterman, C. W. "Status Bonnevie-Ullrich." *Genetics* 33, no. 6 (1948): 607.

Cotterman, C. W., and H. F. Falls. "Unilateral Developmental Anomalies in Sisters." *American Journal of Human Genetics* 1, no. 2 (1949): 203–13.

Cotterman, C. W., H. F. Falls, and J. V. Neel. "Some Hereditary Diseases Having Subclinical Manifestations in Carriers." *Genetics* 33, no. 6 (1948): 608.

Court Brown, W. M. "Sex Chromosomes and the Law." *Lancet* 280, no. 7254 (1962): 508–9.

Couzin-Frankel, J. "Genetics: The Promise of a Cure; 20 Years and Counting." *Science* 324 (2009): 1504–7. doi: 10.1126/science.324_1504.

Cowan, Ruth Schwartz. "Women's Roles in the History of Amniocentesis and Chorionic Villi Sampling." In *Women and Prenatal Testing: Facing the Challenges of Genetic Technology,* ed. Karen H. Rothenberg and Elizabeth Jean Thomson, 35–48. Columbus: Ohio State University Press, 1994.

Creager, Angela N. H. *The Life of a Virus: Tobacco Mosaic Virus as an Experimental Model, 1930–1965.* Chicago: University of Chicago Press, 2002.

Creager, Angela N. H., and G. J. Morgan. "After the Double Helix: Rosalind Franklin's Research on Tobacco Mosaic Virus." *Isis* 99, no. 2 (2008): 239–72.

Crow, James Franklin, and Carter Denniston. "In Memory of Charles W. Cotterman, 1914–89." *American Journal of Human Genetics* 44 (1989): 903–4.

Culley, W. J., E. T. Mertz, M. W. Luce, J. M. Calancro, and J. H. Dolty. "Paper Chromatographic Estimation of Phenylalanine and Tyrosine Using Finger-Tip Blood." *Clinical Chemistry* 8 (1962): 266–69.

Culliton, B. J. "Patients' Rights: Harvard Is Site of Battle over X and Y Chromosomes." *Science* 186 (1974): 715–17. doi: 10.1126/science.186.4165.715.

———. "XYY: Harvard Researcher under Fire Stops Newborn Screening." *Science* 188 (1975): 1284–85. doi: 10.1126/science.11643276.

"The Danger of Medicine Producing Racial Degeneration." *Journal of the American Medical Association* 95, no. 24 (1930): 1844–45.

Darwin, Charles. "Provisional Hypothesis of Pangenesis." In *The Variation of Animals and Plants under Domestication,* ed. Harriet Ritvo, 349–99. 1868; Baltimore: Johns Hopkins University Press, 1998. http://www.esp.org/books/darwin/variation/facsimile/contents/darwin-variation-chap-27-i.pdf.

Davenport, Charles [Benedict]. "Annual Report of the Director of the Department of Experimental Evolution and of the Eugenics Record Office." In *Year Book of the Carnegie Institution of Washington,* 103–26. Washington, DC, 1918.

———. *Eugenics: The Science of Human Improvement by Better Breeding.* New York: Holt, 1910.

———. "Eugenics and the Physician." *New York Medical Journal,* June 8, 1912, 1195–99.

———. "Mendel's Law of Dichotomy in Hybrids." *Biological Bulletin* 2 (1901): 307–10.

———. "The Trait Book." *Eugenics Record Office Bulletin* 6 (1912).

Davies, K[ay]. E., B. D. Young, R. G. Elles, M. E. Hill, and R. Williamson. "Cloning of a Representative Genomic Library of the Human X Chromosome after Sorting by Flow Cytometry." *Nature* 293 (1981): 374–76. doi: 10.1038/293374a0.

Davies, Kevin, and Michael White. *Breakthrough: The Race to Find the Breast Cancer Gene.* New York: Wiley, 1996.

Deichmann, Ute. "A Brief Review of the Early History of Genetics and Its Relationship to Physics and Chemistry." In *Max Delbrück and Cologne: An Early Chapter of German Molecular Biology,* ed. Simone Wenkel and Ute Deichmann, 3–18. Hackensack, NJ: World Scientific, 2007.

———. "Hans Nachtsheim: A Human Geneticist under National Socialism and the Question of Freedom of Science." *Practices of Human Genetics,* ed. Michael Fortun and Everett Mendelsohn, 143–54. Dordrecht: Kluwer, 1999.

Dejong-Lambert, W., and N. Krementsov. "On Labels and Issues: The Lysenko Controversy and the Cold War." *Journal of the History of Biology* (2011). doi: 10.1007/s10739-011-9292-6.

Dent, C. E. "Paper Chromatography and Medicine." In *Recent Advances in Clinical Pathology,* ed. S. C. Dyke, 238–58. Philadelphia: Blakiston, 1951.

de Vries, Hugo. *Die Mutationstheorie.* Leipzig: Veit, 1901.

———. "Sur la loi de disjonction des hybrides." *Comptes Rendus de l'Academie des Sciences* 130 (1900): 845–47.

Dobzhansky, Theodosius. Foreword to Osborn, *The Future of Human Heredity,* v–vii.

Doel, R. E., D. Hoffmann, and N. Krementsov. "National States and International Science: A Comparative History of International Science Congresses in Hitler's Germany, Stalin's Russia, and Cold War United States." *Osiris* 20 (2005): 49–76.

Draper, George. "Biological Philosophy and Medicine." *Human Biology* 1 (1929): 117–35.

———. "The Relationship of Human Constitution to Disease." *Science* 61 (1925): 525–28. doi: 10.1126/science.61.1586.525.

Drouard, Alain. "Concerning Eugenics in Scandinavia. An Evaluation of Recent Research and Publications." *Population: An English Selection* 11 (1999): 261–70.

Drumm, M. L., C. L. Smith, M. Dean, J. L. Cole, M. C. Iannuzzi, and F. S. Collins. "Physical Mapping of the Cystic Fibrosis Region by Pulsed-Field Gel Electrophoresis." *Genomics* 2, no. 4 (1988): 346–54.

Dunn, L. C. "Meeting of Physical Anthropologists and Geneticists for a Definition of the Concept of Race, Paris, 1951." UNESCO report, 1951.

———. *Race and Biology*. Paris: UNESCO, 1951.

Edelson, P. J. "Adopting Osler's Principles: Medical Textbooks in American Medical Schools, 1891–1906." *Bulletin of the History of Medicine* 68, no. 1 (1994): 67–84.

Edwards, J. H., D. G. Harnden, A. H. Cameron, V. M. Crosse, and O. H. Wolff. "A New Trisomic Syndrome." *Lancet* 275, no. 7128 (1960): 787–90.

Emrich, John S. "Dr. Genelove: How Scientists Learned to Stop Worrying and Love Recombinant DNA." Ph.D. diss., George Washington University, 2009.

"The Eugenic Problem." *Journal of the American Medical Association* 100, no. 6 (1933): 435.

Evans, Francis C. "Lee Raymond Dice, 1887–1977." *Journal of Mammalogy* 59, no. 3 (1978): 635–44.

Eyler, John M. *Victorian Social Medicine: The Ideas and Methods of William Farr*. Baltimore: Johns Hopkins University Press, 1979.

Faber, Knud Helge. *Nosography in Modern Internal Medicine*. New York: P. B. Hoeber, 1923.

Falls, Harold F., and Charles W. Cotterman. "Genetic Studies on Ectopia Lentis." *Archives of Ophthalmology* 30 (1943): 610–20.

Falls, H[arold]. F., W. T. Kruse, and C. W. Cotterman. "Three Cases of Marcus Gunn Phenomenon in Two Generations." *American Journal of Ophthalmology* 32, pt. 2, no. 6 (1949): 53–59.

Fangerau, H. " 'Baur-Fischer-Lenz' in 1921–1940 Critical Book Reviews: A Quantitative Study of Contemporary Reception of Racial Eugenics Theories." *Medizinhistorisches Journal* 38, no. 1 (2003): 57–81.

Fangerau, H., and I. Muller. "The Standard Textbook on Racial Hygiene by Erwin Baur, Eugen Fischer, and Fritz Lenz as Viewed by the Psychiatric and Neurological Communities from 1921 to 1940." *Nervenarzt* 73, no. 11 (2002): 1039–46.

Fee, Elizabeth. *Disease and Discovery: A History of the Johns Hopkins School of Hygiene and Public Health, 1916–1939*. Baltimore: Johns Hopkins University Press, 1987.

Feldman, Simon D., and Alfred J. Tauber. "Sickle Cell Anemia: Reexamining the First 'Molecular Disease.' " *Bulletin of the History of Medicine* 71 (1997): 623–50.

Ferguson-Smith, M. A. "Sex Chromatin Anomalies in Mentally Defective Individuals." *Acta Cytologica* 6 (January–February 1962): 73–83.

———. "The Techniques of Human Cytogenetics." *American Journal of Obstetrics and Gynecology* 90 (December 1, 1964): suppl. 1035–54.

Ferguson-Smith, M. A., W. Hamilton, I. C. Ferguson, and P. M. Ellis. "An Abnormal Metacentric Chromosome in an Infant with Leprechaunism." *Annals of Genetics* 11, no. 4 (1968): 195–200.

Ferguson-Smith, M. A., and A. W. Johnston. "Chromosome Abnormalities in Certain Diseases of Man." *Annals of Internal Medicine* 53 (1960): 359–71.

———. "The Human Chromosomes in Disorders of Sex Differentiation." *Transactions of the Association of American Physicians* 73 (1960): 60–71.

Fisher, Irving, D. C. National Conservation Commission Washington, and Committee of One Hundred on National Health. *Bulletin 30 of the Committee of One Hundred on National Health: Being a Report on National Vitality, Its Wastes, and Conservation.* Washington, DC: Government Printing Office, 1909.

Fisher, Irving, Eugene Lyman Fisk, and Life Extension Institute, Incorporated. *How to Live: Rules for Healthful Living, Based on Modern Science.* New York: Funk and Wagnalls, 1915.

Fisher, Ronald A. "Has Mendel's Work Been Rediscovered?" *Annals of Science* 1 (1936): 115–37.

Ford, C. E., and J. L. Hamerton. "The Chromosomes of Man." *Nature* 178 (1956): 1020-23. doi: 10.1038/1781020a0.

Ford, C. E., K. W. Jones, O. J. Miller, U. Mittwoch, L. S. Penrose, M. Ridler, and A. Shapiro. "The Chromosomes in a Patient Showing Both Mongolism and the Klinefelter Syndrome." *Lancet* 273, no. 7075 (1959): 709–10.

Ford, C. E., K. W. Jones, P. E. Polani, J. C. De Almeida, and J. H. Briggs. "A Sex-Chromosome Anomaly in a Case of Gonadal Dysgenesis (Turner's Syndrome)." *Lancet* 273, no. 7075 (1959): 711–13.

Fornek, Scott. "Blind to a Nightmare." *Chicago Sun-Times*, August 31, 2003.

Fox, Arthur L. "The Relationship between Chemical Constitution and Taste." *Proceedings of the National Academy of Sciences USA* 18 (1932): 115–20.

Fraser, F. Clarke. Interview with Andrea Maestrejuan, October 27, 2004, Toronto. Oral History of Human Genetics Project, UCLA and Johns Hopkins University, http://ohhgp.pendari.com.

Friedmann, T., and R. Roblin. "Gene Therapy for Human Genetic Disease?" *Science* 175 (1972): 949–55. doi: 10.1126/science.175.4025.949.

Fuchs, F., E. Freiesleben, E. E. Knudsen, and P. Riis. "Antenatal Detection of Hereditary Diseases." In Kemp, Hauge, and Harvald, *Proceedings of the First International Congress of Human Genetics*, 261–63.

Fuchs, F., and P. Riis. "Antenatal Sex Determination." *Nature* 177 (1956): 330. doi: 10.1038/177330a0.

Galton, Francis. "Eugenics and the Jew (Interview)." *Jewish Chronicle*, July 30, 1910, 16.

———. *Hereditary Genius: An Inquiry into Its Laws and Consequences.* 2nd ed. 1869; London: Macmillan, 1892.

———. "Hereditary Improvement." *Fraser's Magazine* 7 (1873): 116–30.

———. “Hereditary Talent and Character.” *Macmillan’s* 12 (1865): 157–66, 318–27.

———. “The History of Twins, as a Criterion of the Relative Powers of Nature and Nurture.” *Fraser’s Magazine* 12 (1875): 566–76.

———. *Inquiries into Human Faculty and Its Development,* 1883; 2nd ed. London: J. M. Dent, 1907.

———. *Natural Inheritance.* London: Macmillan, 1889.

———. “The Possible Improvement of the Human Breed under Existing Conditions of Law and Sentiment.” In *Essays in Eugenics,* 1–34. London: Eugenics Education Society, 1909.

———. “The Possible Improvement of the Human Breed under the Existing Conditions of Law and Sentiment.” *Nature* 64 (1901): 659–65. doi: 10.1038/064659b0.

———. “The Relative Sensitivity of Men and Women at the Nape of the Neck, by Weber’s Test.” *Nature* 50 (1894): 40–42. doi: 10.1038/050040a0.

———. “Statistical Inquiries into the Efficacy of Prayer.” *Fortnightly Review* 12 (1872): 125–35.

———. “A Theory of Heredity.” *Journal of the Anthropological Institute* 5 (1876): 329–48.

Garrod, Archibald Edward. “About Alkaptonuria.” *Lancet* 158 (1901): 1484–86.

———. *Inborn Errors of Metabolism: The Croonian Lectures Delivered before the Royal College of Physicians of London in June 1908.* London: H. Frowde and Hodder & Stoughton, 1909.

———. “The Incidence of Alkaptonuria: A Study in Chemical Individuality.” *Lancet* 160, no. 4137 (1902): 1616–20.

Garrod, Archibald Edward, Charles R. Scriver, and Barton Childs. *Garrod’s Inborn Factors in Disease: Including an Annotated Facsimile Reprint of the Inborn Factors in Disease by Archibald E. Garrod.* Oxford Medical Publications. Oxford: Oxford University Press, 1989.

Gaudillière, J. P. “Making Heredity in Mice and Men: The Production and Uses of Animal Models in Postwar Human Genetics.” In *Heredity and Infection: The History of Disease Transmission,* ed. Jean-Paul Gaudillière and Ilana Löwy, 181–202. London: Routledge, 2001.

———. “Mendelism and Medicine: Controlling Human Inheritance in Local Contexts, 1920–1960.” *Comptes Rendus de l’Académie des Sciences,* series 3, *Sciences de la vie* 323, no. 12 (2000): 1117–26.

“Genetic Effects of the Atomic Bombs in Hiroshima and Nagasaki.” *Science* 106 (1947): 331–33. doi: 10.1126/science.106.2754.331.

Gershon, E. S. “The Historical Context of Franz Kallmann and Psychiatric Genetics.” *Archiv für Psychiatrie und Nervenkrankheiten* 229, no. 4 (1981): 273–76.

Gillham, Nicholas W. *A Life of Sir Francis Galton: From African Exploration to the Birth of Eugenics.* New York: Oxford University Press, 2001.

———. “Sir Francis Galton and the Birth of Eugenics.” *Annual Review of Genetics* 35 (2001): 83–101.

Gingras, Yves. "Revisiting the 'Quiet Debut' of the Double Helix: A Bibliometric and Methodological Note on the 'Impact' of Scientific Publications." *Journal of the History of Biology* 43, no. 1 (2010): 159–81.

Glass, Bentley. "Genetics in the Service of Man." *Johns Hopkins Magazine* 6 (1955): 2–5, 12–16.

Glickman, Lawrence B., and Steven J. Diner. *A Very Different Age: Americans of the Progressive Era*. New York: Hill and Wang, 1998.

Gormley, M. "Scientific Discrimination and the Activist Scientist: L. C. Dunn and the Professionalization of Genetics and Human Genetics in the United States." *Journal of the History of Biology* 42 (2009): 33–72.

Gould, Stephen Jay. *Mismeasure of Man*. New York: Norton, 1996.

Greene, Jeremy A. *Prescribing by Numbers: Drugs and the Definition of Disease*. Baltimore: Johns Hopkins University Press, 2007.

Greenwald, Brian H. "Alexander Graham Bell through the Lens of Eugenics, 1883—1922." Ph.D. diss., George Washington University, 2006.

Groeben, Christiane, Jane Maienschein, Alberto Monroy, Nathan Reingold, Joel N. Bodansky, Paul R. Gross, James D. Ebert, et al. "The Naples Zoological Station and the Marine Biological Laboratory: One Hundred Years of Biology." *Biological Bulletin* 168 (1985 supplement): 1–207.

Gruenberg, H. "Men and Mice at Edinburgh: Reports from the Genetics Congress." *Journal of Heredity* 30 (1939): 371–74.

Guterman, Lila. "Gene-Therapy Pioneer Convicted of Abuse." *Chronicle of Higher Education* 52, no. 48 (2006): A10.

Guthrie, Robert. "Blood Screening for Phenylketonuria." *JAMA* 178 (1961): 863.

Guthrie, R[obert]., and A. Susi. "A Simple Phenylalanine Method for Detecting Phenylketonuria in Large Populations of Newborn Infants." *Pediatrics* 32 (1963): 338–43.

Guyer, M. F. "The Germinal Background of Somatic Modifications." *Science* 71 (1930): 169–76. doi: 10.1126/science.71.1833.169.

———. "The Internal Secretions and Human Well-Being." *Science* 74 (1931): 159–66. doi: 10.1126/science.74.1911.159.

———. "A Note on the Accessory Chromosomes of Man." *Science* 39 (May 15, 1914): 721–22. doi: 10.1126/science.39.1011.721.

Guyer, M. F., and E. A. Smith. "Transmission of Eye-Defects Induced in Rabbits by Means of Lens-Sensitized Fowl-Serum." *Proceedings of the National Academy of Sciences U S A* 6, no. 3 (1920): 134–36.

Haldane, J. B. S. *The Biochemistry of Genetics*. London: Allen and Unwin, 1952.

———. "Some Alternatives to Sex." *New Biology* 19 (1955): 7–26.

———. "William Bateson." In *Possible Worlds*, ed. Carl A. Price, 135–39. New Brunswick, NJ: Transaction, 2000.

Harden, Victoria Angela. *Inventing the NIH: Federal Biomedical Research Policy, 1887–1937*. Baltimore: Johns Hopkins University Press, 1986.

Harper, Peter S. *First Years of Human Chromosomes: The Beginnings of Human Cytogenetics*. Bloxham: Scion, 2006.

———. *A Short History of Medical Genetics*. Oxford Monographs on Medical Genetics. Oxford: Oxford University Press, 2008.

———. "William Bateson, Human Genetics and Medicine." *Human Genetics* 118, no. 1 (2005): 141–51.

Harper, P[eter]. S., T. O'Brien, J. M. Murray, K. E. Davies, P. Pearson, and R. Williamson. "The Use of Linked DNA Polymorphisms for Genotype Prediction in Families with Duchenne Muscular Dystrophy." *Journal of Medical Genetics* 20, no. 4 (1983): 252–54.

Harris, Harry. "Enzyme Polymorphisms in Man." *Proceedings of the Royal Society of London B, Biological Sciences* 164, no. 995 (1966): 298–310.

———. "The 'Inborn Errors' Today." In *Garrod's Inborn Errors of Metabolism*, 120–83. London: Oxford University Press, 1963.

———. "Lionel Sharples Penrose, 1898–1972." *Biographical Memoirs of the Fellows of the Royal Society* 19 (1973): 521–61.

Harris, Henry. *The Cells of the Body: A History of Somatic Cell Genetics*. Cold Spring Harbor, NY: Cold Spring Harbor Laboratory Press, 1995.

Harris, John. *Enhancing Evolution: The Ethical Case for Making Better People*. Princeton: Princeton University Press, 2007.

Harvey, A. McGehee, Victor A. McKusick, and John D. Stobo. *Osler's Legacy: The Department of Medicine at Johns Hopkins, 1889–1989*. Baltimore: Department of Medicine, Johns Hopkins University, 1990.

Harvey, A. M., G. H. Brieger, S. L. Abrams, and V. A. McKusick. "A Model of Its Kind: A Century of Medicine at Johns Hopkins." *Journal of the American Medical Association* 261, no. 21 (1989): 3136–42.

Hayes, Brian. "The Invention of the Genetic Code." *American Scientist* 86, no. 1 (1998): 8–14.

Herndon, Claude Nash. Interview with Elizabeth Allan Berger, 1976. Claude Nash Herndon papers, Coy C. Carpenter Library, Wake Forest University.

———. "William Allan: An Appreciation." *American Journal of Human Genetics* 14 (1962): 97–101.

Hiltzik, Lee-Richard. "The Brooklyn Institute of Arts and Sciences' Biological Laboratory, 1890–1924: A History." Ph.D. diss., State University of New York, 1993.

Hirschhorn, Kurt. Interview with Andrea Maestrejuan, August 7, 2002, New York. Oral History of Human Genetics Project, UCLA and Johns Hopkins University, http://ohhgp.pendari.com.

Hirshbein, Laura Davidow. "Masculinity, Work, and the Fountain of Youth: Irving Fisher and the Life Extension Institute, 1914–31." *Canadian Bulletin of Medical History* 16, no. 1 (1999): 89–124.

Hogben, Lancelot. *Nature and Nurture*. London: Allen and Unwin, 1933.

Holmes, Frederic L. *Meselson, Stahl, and the Replication of DNA: The History of "the Most Beautiful Experiment in the World."* New Haven: Yale University Press, 2001.

Hood, Leroy. "Biology and Medicine in the Twenty-First Century." In *The Code of Codes: Scientific and Social Issues in the Human Genome Project,* ed. D. J. Kevles and L. Hood, 136–64. Cambridge: Harvard University Press, 1992.

Hooton, Earnest A. "Human Heredity or Forbidden Fruit of the Tree of Knowledge." In *Medical Genetics and Eugenics,* 40–60. Philadelphia: Woman's Medical College of Pennsylvania, 1943.

Hsu, T. C. *Human and Mammalian Cytogenetics: A Historical Perspective.* New York: Springer-Verlag, 1979.

———. "Mammalian Chromosomes in Vitro: I. The Karyotype of Man." *Journal of Heredity* 43 (1952): 167–72.

"Indications for Abortion and the Law." *Journal of the American Medical Association* 96, no. 21 (1931): 1810–11.

Ingram, Vernon M. "Gene Mutations in Human Haemoglobin: The Chemical Difference between Normal and Sickle Cell Haemoglobin." *Nature* 180 (1957): 326–28. doi: 10.1038/180326a0.

"Interim Committee of International Eugenics Congress." *Eugenical News* 6, no. 11 (1921): 65–67.

Jacob, François, D. Perrin, C. Sanchez, and Jacques Monod. "L'opéron: Groupe de gènes à expresion coordonnée par un opérateur." *Comptes Rendus de l'Académie des Sciences* 250 (1960): 1727–29.

Jacobs, P[atricia]. A., A. G. Baikie, W. M. Court Brown, H. Forrest, J. R. Roy, J. S. Stewart, and B. Lennox. "Chromosomal Sex in the Syndrome of Testicular Feminisation." *Lancet* 274, no. 7103 (1959): 591–92.

Jacobs, P. A., A. G. Baikie, W. M. Court Brown, T. N. Macgregor, N. Maclean, and D. G. Harnden. "Evidence for the Existence of the Human 'Super Female.'" *Lancet* 274, no. 7100 (1959): 423–25.

Jacobs, P. A., A. G. Baikie, W. M. Court Brown, and J. A. Strong. "The Somatic Chromosomes in Mongolism." *Lancet* 273, no. 7075 (1959): 710.

Jacobs, Patricia A., M. Brunton, M. M. Melville, R. P. Brittain, and W. F. McClemont. "Aggressive Behavior, Mental Subnormality, and the XYY Male." *Nature* 208 (1965): 1351–52. doi: 10.1038/2081351a0.

Jacobs, Patricia A., and J. A. Strong. "A Case of Human Intersexuality Having a Possible XXY Sex-Determining Mechanism." *Nature* 183 (1959): 302–3. doi: 10.1038/183302a0.

Jeghers, Harold, Victor A. McKusick, and K. H. Katz. "Generalized Intestinal Polyposis and Melanin Spots of the Oral Mucosa, Lips, and Digits: A Syndrome of Diagnostic Significance." *New England Journal of Medicine* 241, no. 25 (1949): 993–1005.

Jennings, Herbert Spencer. "Raymond Pearl (1879–1940)." *Biographical Memoirs of the National Academy of Sciences* 22 (1942): 294–347.

Johnston, A. W., M. A. Ferguson-Smith, S. D. Handmaker, H. W. Jones, and G. S. Jones. "The Triple-X Syndrome: Clinical, Pathological, and Chromosomal Studies in Three Mentally Retarded Cases." *British Medical Journal* 2, no. 5259 (1961): 1046–52.

Jordan, H. E. "The Place of Eugenics in the Medical Curriculum." In *Problems in Eugenics: Papers Communicated to the First International Eugenics Congress*, 396–99. London: Eugenics Education Society, 1912.

Judson, Horace Freeland. *The Eighth Day of Creation: Makers of the Revolution in Biology*, 2nd ed. Cold Spring Harbor, NY: Cold Spring Harbor Laboratory Press, 1996.

Kaiser, J. "Gene Therapy: Beta-Thalassemia Treatment Succeeds, with a Caveat." *Science* 326, (2009): 1468–69. doi: 10.1126/science.326.5959.1468-b.

Kamrat-Lang, Debora. "Healing Society: Medical Language in American Eugenics." *Science in Context* 8, no. 1 (1995): 175–96.

Kan, Y. W., and A. M. Dozy. "Polymorphism of DNA Sequence Adjacent to Human Beta-Globin Structural Gene: Relationship to Sickle Mutation." *Proceedings of the National Academy of Sciences U S A* 75, no. 11 (1978): 5631–35.

Kay, Lily E. "Laboratory Technology and Biological Knowledge: The Tiselius Electrophoresis Apparatus, 1930–1945." *History and Philosophy of the Life Sciences* 10, no. 1 (1988): 51–72.

———. *The Molecular Vision of Life: Caltech, the Rockefeller Foundation, and the Rise of the New Biology*. New York: Oxford University Press, 1993.

———. *Who Wrote the Book of Life?* Stanford, CA: Stanford University Press, 2000.

Keating, Peter. "Holistic Bacteriology: Ludvik Hirszfeld's Doctrine of Serogenesis between the Two World Wars." In Lawrence and Weisz, *Greater than the Parts*, 283–302.

Keller, Evelyn Fox. *The Century of the Gene*. Cambridge: Harvard University Press, 2001.

Kellogg, John Harvey. *Plain Facts for Old and Young*. Burlington, IA: I. F. Segner, 1881.

Kemp, Tage. "Address at the Opening of the First International Congress of Human Genetics, Aug. 1st, 1956." In Kemp, Hauge, and Harvald, *Proceedings of the First International Congress of Human Genetics*, xii–xiii.

Kemp, Tage, Mogens Hauge, and Bent Harvald. *Proceedings of the First International Congress of Human Genetics, Copenhagen, August 1–6, 1956*. Basel: Karger, 1957.

Kendrick, Brig. Gen. Douglas B. *Blood Program in World War II*. Washington, DC: Office of the Surgeon General, Department of the Army, Government Printing Office, 1964.

Kerem, B., J. M. Rommens, J. A. Buchanan, D. Markiewicz, T. K. Cox, A. Chakravarti, M. Buchwald, and L. C. Tsui. "Identification of the Cystic Fibrosis Gene: Genetic Analysis." *Science* 245 (1989): 1073–80. doi: 10.1126/science.2570460.

Kevles, Daniel J. *In the Name of Eugenics: Genetics and the Uses of Human Heredity*. 2nd ed. Cambridge: Harvard University Press, 1995.

———. "The National Science Foundation and the Debate over Postwar Research Policy, 1942–1945: A Political Interpretation of Science—the Endless Frontier." *Isis* 68, no. 241 (1977): 5–26.

Kimmelman, Barbara A. "The American Breeders' Association: Genetics and Eugenics in an Agricultural Context, 1903–13." *Social Studies of Science* 13 (1983): 163–204.

———. "A Progressive Era Discipline: Genetics at American Agricultural Colleges and Experiment Stations, 1900–1920." Ph.D. diss., University of Pennsylvania, 1987.

Kluchin, Rebecca M. *Fit to Be Tied: Sterilization and Reproductive Rights in America, 1950–1980*. Critical Issues in Health and Medicine. New Brunswick, NJ: Rutgers University Press, 2009.

Knopf, S. Adolphus. "A Federal Department of Health." *Popular Science Monthly* 77 (July–December 1910): 373–78.

Koch, Lene. "Eugenic Sterilisation in Scandinavia." *European Legacy* 11, no. 3 (2006): 299–309.

Koenig, M., E. P. Hoffman, C. J. Bertelson, A. P. Monaco, C. Feener, and L. M. Kunkel. "Complete Cloning of the Duchenne Muscular Dystrophy (Dmd) Cdna and Preliminary Genomic Organization of the Dmd Gene in Normal and Affected Individuals." *Cell* 50, no. 3 (1987): 509–17. doi: 0092-8674(87)90504-6.

Kohler, Robert. *Lords of the Fly: Drosophila Genetics and the Experimental Life*. Chicago: University of Chicago Press, 1994.

Kolb, Lawrence C., and Leon Roizin. *The First Psychiatric Institute: How Research and Education Changed Practice*. Washington, DC: American Psychiatric Press, 1993.

Kottler, M. J. "From 48 to 46: Cytological Technique, Preconception, and the Counting of Human Chromosomes." *Bulletin of the History of Medicine* 48, no. 4 (1974): 465–502.

Krementsov, N. "From 'Beastly Philosophy' to Medical Genetics: Eugenics in Russia and the Soviet Union." *Annals of Science* 68, no. 1 (2011): 61–92.

———. *International Science between the World Wars: The Case of Genetics*. London: Routledge, 2005.

———. "A 'Second Front' in Soviet Genetics: The International Dimension of the Lysenko Controversy, 1944–1947." *Journal of the History of Biology* 29, no. 2 (1996): 229–50.

Krimsky, Sheldon. *Genetic Alchemy: The Social History of the Recombinant DNA Controversy*. Cambridge: MIT Press, 1982.

Kunkel, H. G., and R. J. Slater. "Zone Electrophoresis in a Starch Supporting Medium." *Proceedings of the Society for Experimental Biology and Medicine* 80, no. 1 (1952): 42–44.

Kuska, B. "Beer, Bethesda, and Biology: How 'Genomics' Came into Being." *Journal of the National Cancer Institute* 90, no. 2 (1998): 93.

Lamb, Susan. "Adolf Meyer: Pathologist of the Mind." Ph.D. diss., Johns Hopkins University, 2010.

Landecker, Hannah. *Culturing Life: How Cells Became Technologies*. Cambridge: Harvard University Press, 2007.

Lander, E. S., and N. J. Schork. "Genetic Dissection of Complex Traits." *Science* 265 (1994): 2037–48. doi: 10.1126/science.8091226.

Laughlin, Harry. "Studies in Eugenics and Heredity." In *Annual Report of the Director of the Department of Genetics*, ed. Albert F. Blakeslee, 60–66. Cold Spring Harbor, NY: Carnegie Institute of Washington, 1938.

Laurence, William L. "Nobel Winners: Importance of the Discovery of the Structure of DNA Is Examined." *New York Times,* October 21, 1962.

Lawrence, Christopher, and George Weisz, eds. *Greater than the Parts: Holism in Biomedicine, 1920–1950.* New York: Oxford University Press, 1998.

Laxova, R. "Lionel Sharples Penrose, 1898–1972: A Personal Memoir in Celebration of the Centenary of His Birth." *Genetics* 150 (1998): 1333–40.

Lederberg, Joshua. "Chromosomes and Crime." *Washington Post,* January 22, 1.

———. "Experimental Genetics and Human Evolution." *American Naturalist* 100, no. 915 (1966): 519–31.

———. Interviews with Nathaniel Comfort, March 12, May 26, 1996, New York. American Philosophical Society Library.

———. "Memorandum to Dean Bowers." In Joshua Lederberg Papers, National Library of Medicine, Bethesda, MD, 1955.

———. "Molecular Biology, Eugenics, and Euphenics." *Nature* 198 (1963): 428–29, doi: 10.1038/198428a0.

———. "Some Early Stirrings (1950 ff.) of Concern about Environmental Mutagens." *Environmental and Molecular Mutagenesis* 30, no. 1 (1997): 3–10.

Lederberg, Joshua, and Edward Lawrie Tatum. "Gene Recombination in Escherichia coli." *Nature* 158 (1946): 558. doi: 10.1038/158558a0.

Lederman, Muriel, Richard M. Burian, Sue A. Tolin, William C. Summers, Doris T. Zallen, Robert E. Kohler, Frederic L. Holmes, and Bonnie Tocher Clause. "The Right Organism for the Job." *Journal of the History of Biology* 26 (1993): 233–367.

Lejeune, Jerome, Marthe Gautier, and Raymond Turpin. "Etude des chromosomes somatiques de neuf enfants mongoliens" (Study of the somatic chromosomes of nine Mongoloid idiot children). *Comptes Rendus de l'Académie de Sciences* 248 (1959): 1721–22.

Lejeune, J[erome]., R. Turpin, and M. Gautier. "Le mongolisme, premier exemple d'aberration autosomique humaine." *Annals of Genetics* 1, no. 2 (1959): 41–49.

Lindee, Susan. "Babies' Blood." In *Moments of Truth in Genetic Medicine,* 28–57.

———. "Genetic Disease in the 1960s: A Structural Revolution." *American Journal of Medical Genetics* 115, no. 2 (2002): 75–82.

———. *Moments of Truth in Genetic Medicine.* Baltimore: Johns Hopkins University Press, 2005.

———. "Provenance and the Pedigree: Victor McKusick's Fieldwork with Ellis Van Creveld Syndrome in the Pennsylvania Amish." In *Genetic Nature/Culture: Anthropology and Science beyond the Two-Culture Divide.* Berkeley: University of California Press, 2003.

———. "Squashed Spiders." In *Moments of Truth in Genetic Medicine,* 90–119.

———. *Suffering Made Real: American Science and the Survivors at Hiroshima.* Chicago: University of Chicago Press, 1994.

"List of the American Passengers aboard the Torpedoed Athenia." *New York Times,* September 4, 1939.

Little, Clarence Cook. "Opportunities for Research in Mammalian Genetics." *Scientific Monthly* 26, no. 6 (1928): 521–34.

Lyon, Jeff, and Peter Gorner. *Altered Fates: Gene Therapy and the Retooling of Human Life*. New York: Norton, 1995.

Lyon, M. F. "Gene Action in the X-Chromosome of the Mouse (Mus Musculus L.)." *Nature* 190 (1961): 372–73. doi: 10.1038/190372a0.

MacDowell, E. Carleton. "Charles Benedict Davenport, 1886–1944: A Study of Conflicting Influences." *Bios* 70 (1946): 3–50.

MacKenzie, David, and Barry Barnes. "Scientific Judgment: The Biometry—Mendelian Controversy." In *Natural Order: Historical Studies of Scientific Culture*, ed. Barry Barnes and Steven Shapin, 191–210. Beverly Hills, CA: Sage, 1979.

Macklin, Madge Thurlow. "Hereditary Abnormalities of the Eye: I. Introduction: The Laws of Heredity and Their Exemplification in the Inheritance of Eye Color." *Canadian Medical Association Journal* 16 (1926): 1340–42.

———. "Hereditary Abnormalities of the Eye: III. (Section 2) Anomalies of the Entire Eyeball (Continued)." *Canadian Medical Association Journal* 17 (1927): 421–23.

———. "'Medical Genetics': A Necessity in the Up-to-Date Medical Curriculum." *Journal of Heredity* 23 (1932): 485–86.

———. "Medical Genetics: An Essential Part of the Medical Curriculum from the Standpoint of Prevention." *Journal of the Association of American Medical Colleges* 8 (1933): 291–304.

———. "Should the Teaching of Genetics as Applied to Medicine Have a Place in the Medical Curriculum?" *Journal of the Association of American Medical Colleges* 7 (1932): 368–73.

———. "The Teaching of Inheritance of Disease to Medical Students: A Proposed Course in Medical Genetics." *Annals of Internal Medicine* 6 (1933): 1335–43.

Macklin, Madge Thurlow, and J. Thornley Bowman. "Inheritance of Peroneal Atrophy." *Journal of the American Medical Association* 86, no. 9 (1926): 613–16.

Maclean, N., J. M. Mitchell, D. G. Harnden, Jane Williams, Patricia A. Jacobs, Karin A. Buckton, A. G. Baikie, et al. "A Survey of Sex-Chromosome Abnormalities among 4514 Mental Defectives." *Lancet* 279, no. 7224 (1962): 293–96.

Maddox, Brenda. *Rosalind Franklin: The Dark Lady of DNA*. New York: HarperCollins, 2002.

Maienschein, Jane. *Defining Biology: Lectures from the 1890s*. Cambridge: Harvard University Press, 1986.

———. "'It's a Long Way from Amphioxus': Anton Dohrn and Late Nineteenth Century Debates about Vertebrate Origins." *History and Philosophy of the Life Sciences*, no. 3 (1994): 465–78.

———. *100 Years of Exploring Life, 1888–1988: The Marine Biological Laboratory at Woods Hole*. Boston: Jones and Bartlett, 1989.

———. *Transforming Traditions in American Biology, 1880–1915*. Baltimore: Johns Hopkins University Press, 1991.

Martin, Aryn. "Can't Any Body Count? Counting as an Epistemic Theme in the History of Human Chromosomes." *Social Studies of Science* 34, no. 6 (2004): 923–48. doi: 10.1177/0306312704046843.

Matthaei, J. Heinrich, and Marshall W. Nirenberg. "Characteristics and Stabilization of DNAase-Sensitive Protein Synthesis in E. coli Extracts." *Proceedings of the National Academy of Sciences U S A* 47 (1961): 1580–88.

Mazumdar, Pauline M. H. "Two Models for Human Genetics: Blood Grouping and Psychiatry in Germany between the World Wars." *Bulletin of the History of Medicine* 70, no. 4 (1996): 609–57.

McCabe, Linda L., and E. R. McCabe. "Are We Entering a 'Perfect Storm' for a Resurgence of Eugenics? Science, Medicine, and Their Social Context." In *A Century of Eugenics in America*, ed. Paul Lombardo, 193–218. Bloomington: Indiana University Press, 2011.

McElheny, Victor K. *Watson and DNA: Making a Scientific Revolution.* Cambridge, MA: Perseus, 2003.

McKeown, Thomas. *Modern Rise of Population.* New York: Academic Press, 1976.

McKusick, Victor A. "Birth Defects: Prospects for Progress." In *Congenital Malformations: Proceedings of the Third International Conference, the Hague, the Netherlands, 7–13 Sept. 1969*, ed. F. Clarke Fraser and Victor A. McKusick. Amsterdam: Excerpta Medica, 1970.

———. "The Growth and Development of Human Genetics as a Clinical Discipline." *American Journal of Human Genetics* 27, no. 3 (May 1975): 261–73.

———. *Heritable Disorders of Connective Tissue.* St. Louis: Mosby, 1956.

———. "The Human Gene Map, 20 October 1982." *Clinical Genetics* 22, no. 6 (1982): 360–91.

———. Interview with Andrea Maestrejuan, December 10, 2001, Baltimore. Oral History of Human Genetics Project, UCLA and Johns Hopkins University, http://ohhgp.pendari.com.

———. "Medical Genetics, 1962." *Journal of Chronic Diseases* 16 (1963): 457–634.

———. "Mendelian Inheritance in Man and Its Online Version, OMIM." *American Journal of Human Genetics* 80, no. 4 (2007): 588–604. doi: 10.1086/514346.

———. "On Lumpers and Splitters, or the Nosology of Genetic Disease." *Perspectives in Biology and Medicine* 12, no. 2 (1969): 298–312.

———. "Persisting Memories of Cyril Clarke in Baltimore." *Journal of Medical Genetics* 38 (2001): 284.

———. "A 60-Year Tale of Spots, Maps, and Genes." *Annual Review of Genomics and Human Genetics* 7 (2006): 1–27.

McKusick, Victor A., and Frank H. Ruddle. "The Status of the Gene Map of the Human Chromosomes." *Science* 196 (1977): 390–405, doi: 10.1126/science.850784.

McKusick, Vincent L. Interview with Andrea L'Hommedieu, July 23, 2001, Portland, Maine, http://digilib.bates.edu/collect/muskieor/index/assoc/HASH0178/2b0453b3.dir/doc.pdf.

Meads, Manson, Bowman Gray School of Medicine, and North Carolina Baptist Hospital. *The Miracle on Hawthorne Hill: A History of the Medical Center of the Bowman Gray School of Medicine of Wake Forest University and the North Carolina Baptist Hospital.* Winston-Salem, NC: Wake Forest University, 1988.

Mehler, Barry. "A History of the American Eugenics Society, 1921–1940." Ph.D. diss., University of Illinois, 1988.

———. "Macklin, Madge Thurlow, Feb. 6, 1893–March 14, 1962." In *Notable American Women: The Modern Period; A Biographical Dictionary,* ed. Barbara Sicherman and Carol Hurd Green, with Ilene Kantrov and Harriette Walker, xxii. Cambridge: Belknap Press of Harvard University Press, 1980.

Mehlman, Maxwell J. *The Price of Perfection: Individualism and Society in the Era of Biomedical Enhancement.* Baltimore: Johns Hopkins University Press, 2009.

Mendelsohn, J. Andrew. "Medicine and the Making of Bodily Inequality in Twentieth-Century Europe." In *Heredity and Infection: The History of Disease Transmission,* ed. Jean-Paul Gaudilliere and Ilana Löwy, 21–79. London: Routledge, 2001.

Migeon, Barbara R. Interview with Jennifer Caron, June 2–3, 2005, Baltimore. Oral history of Human Genetics Project, UCLA and Johns Hopkins University, http://ohhgp.pendari.com/.

Mildenberger, F. "On the Track of 'Scientific Pursuit': Franz Josef Kallmann (1897–1965) and Genetic Racial Research." *Medizinhistorisches Journal* 37, no. 2 (2002): 183–200.

Monaghan, Floyd V., and Alain F. Corcos. "On the Origins of the Mendelian Laws." *Journal of Heredity* 75 (1984): 67–69.

"Mongolian Idiocy." *Journal of the American Medical Association* 95, no. 9 (1930): 684.

Moss, A. J. "Introductory Note to a Classic Article by Francis Galton." *Annals of Noninvasive Electrocardiology* 8, no. 2 (2003): 170.

Motulsky, Arno G. "Drug Reactions, Enzymes, and Biochemical Genetics." *JAMA* 165 (1957): 835–37.

———. Interview with Andrea Maestrejuan, September 4, 2002, Seattle. Oral History of Human Genetics Project, UCLA and Johns Hopkins University.

Mueller-Hill, Benno. *The Lac Operon: A Short History of a Genetic Paradigm.* Berlin: de Gruyter, 1996.

Muller, Hermann Joseph. "Mutation." In *Eugenics, Genetics, and the Family, Scientific Papers of the Second International Congress of Eugenics,* 106–12. Baltimore: Williams and Wilkins, 1923.

———. "Our Load of Mutations." *American Journal of Human Genetics* 2, no. 2 (1950): 111–76.

———. *Out of the Night: A Biologist's View of the Future.* New York: Vanguard, 1935.

———. "Pilgrim Trust Lecture: The Gene." *Proceedings of the Royal Society of London,* Series B, *Biological Sciences* 134, no. 874 (1947): 1–37.

———. "The Production of Mutations." *Journal of Heredity* 38, no. 9 (1947): 258–70.

———. "Progress and Prospects in Human Genetics." *American Journal of Human Genetics* 1, no. 1 (1949): 1–18.

———. "Social Biology and Population Improvement." *Nature* 144 (1939): 521–22. doi: 10.1038/144521a0.

Murlin, J. R. "Science and Culture." *Science* 80 (1934): 81–86. doi: 10.1126/science.80.2065.81.

Murray, J. M., K. E. Davies, P. S. Harper, L. Meredith, C. R. Mueller, and R. Williamson. "Linkage Relationship of a Cloned DNA Sequence on the Short Arm of the X Chromosome to Duchenne Muscular Dystrophy." *Nature* 300 (1982): 69–71. doi: 10.1038/300069a0.

Nagy, G. K. "Sir Francis Galton and Proficiency Testing in Cytopathology." *Acta Cytologica* 51, no. 4 (2007): 530–32.

Naldini, L. "Medicine: A Comeback for Gene Therapy." *Science* 326 (2009): 805–6. doi: 10.1126/science.1181937.

Neel, J[ames]. V. "The Detection of the Genetic Carriers of Hereditary Disease." *American Journal of Human Genetics* 1, no. 1 (1949): 19–36.

———. "The Inheritance of Sickle Cell Anemia." *Science* 110 (1949): 64. doi: 10.1126/science.110.2846.64.

———. "Medicine's Genetic Horizons." *Annals of Internal Medicine* 49 (1958): 472–76.

———. *Physician to the Gene Pool: Genetic Lessons and Other Stories.* New York: Wiley, 1994.

Nelkin, Dorothy, and Susan Lindee. *The DNA Mystique: The Gene as a Cultural Icon.* New York: Freeman, 1995.

Newman, H[oratio]. H[ackett]. "Mental and Physical Traits of Identical Twins Reared Apart: Case I. Twins 'A' and 'O.'" *Journal of Heredity* 29, no. 2 (1929): 49–64.

Newman, Horatio Hackett, Frank Nugent Freeman, and Karl John Holzinger. *Twins: A Study of Heredity and Environment.* Chicago: University of Chicago Press, 1937.

New York (State) Board of Charities, Dept. of State and Alien Poor. *Field Work Manual.* Eugenics and Social Welfare Bulletin 10. Albany, NY: The Capitol, 1917.

Nirenberg, Marshall W., and J. Heinrich Matthaei. "The Dependence of Cell-Free Protein Synthesis in E. Coli upon Naturally Occurring or Synthetic Polyribonucleotides." *Proceedings of the National Academy of Sciences U S A* 47 (1961): 1588–1602.

Norton, J. Pease. "Economic Advisability of a National Department of Health." *Science* 24 (1906): 519–20. doi: 10.1126/science.24.617.519.

Nowell, Peter C. Interview with Nathaniel Comfort, June 27, 2007, Philadelphia. Oral History of Human Genetics Project, UCLA and Johns Hopkins University.

Nowell, P[eter]. C., and D. A. Hungerford. "Aetiology of Leukaemia." *Lancet* 275, no. 7115 (1960): 113–14.

———. "Chromosome Studies on Normal and Leukemic Human Leukocytes." *Journal of the National Cancer Institute* 25, no. 1 (1960): 85–109.

———. "Minute Chromosome in Human Chronic Granulocytic Leukemia." *Science* 132 (1960): 1497.

Nowell, Peter C., David A. Hungerford, and C. D. Brooks. "Chromosomal Characteristics of Normal and Leukemic Human Leukocytes after Short-Term Culture." *Proceedings of the American Association of Cancer Research* 2 (1958): 331–32.

Office of Technology Assessment, United States Congress. *Technology Transfer at the National Institutes of Health.* Washington, DC: Government Printing Office, 1982.

Olby, Robert. *Origins of Mendelism.* Chicago: University of Chicago Press, 1985.

———. "Quiet Debut for the Double Helix." *Nature* 421 (2003): 402–5. doi: 10.1038/nature01397.

Oliver, Robert. *Making the Modern Medical School: The Wisconsin Stories.* Canton, MA: Science History Publications, 2002.

Olson, Maynard. "Davenport's Dream." In Witkowski, Inglis, and Davenport, *Davenport's Dream,* 77–98.

"On the Permanent Commission of Blood Group Investigation." *Eugenical News* 13, no. 6 (1928): 79.

Opitz, J. M. "Biographical Note—Laurence H. Snyder." *American Journal of Medical Genetics* 8, no. 4 (1981): 447–48.

———. "Obituary: Charles W. Cotterman." *American Journal of Medical Genetics* 34, no. 2 (1989): 149–54.

Osborn, Frederick Henry. *The Future of Human Heredity: An Introduction to Eugenics in Modern Society.* New York: Weybright and Talley, 1968.

———. "History of the American Eugenics Society." *Social Biology* 21, no. 2 (1974): 115–26.

———. *Preface to Eugenics.* New York, London: Harper and Brothers, 1940.

Osmundsen, John. "Biologists Hopeful of Solving Secrets of Heredity This Year: Wide Gain Likely in Genetic Study; Science Hopeful of Finding Basis of Thought and Cures for Some Inherited Ills . . . " *New York Times,* February 2, 1962.

Patau, K., D. W. Snuth, E. Therman, S. L. Inhorn, and H. P. Wagner. "Multiple Congenital Anomaly Caused by an Extra Autosome." *Lancet* 275, no. 7128 (1960): 790–93.

Paton, Stewart. "Medicine's Opportunity." *Eugenics in Race and State: Scientific Papers of the Second International Congress of Eugenics* 2 (1923): 327–29.

Paul, Diane B. "Did Eugenics Rest on an Elementary Mistake?" In *Politics of Heredity,* 117–32.

———. "Eugenics and the Left." *Journal of the History of Ideas* 45 (1984): 567–90.

———. "Genes and Contagious Disease: The Rise and Fall of a Metaphor." In *Politics of Heredity,* 142–54.

———. "The History of Newborn Screening for Phenylketonuria in the U. S." In *Promoting Safe and Effective Genetic Testing in the United States: Final Report of the Task Force on Genetic Testing,* 137–60. Baltimore: Johns Hopkins University Press, 1998, http://www.genome.gov/10002397.

———. "Is Human Genetics Disguised Eugenics?" In *Genes and Human Self-Knowledge: Historical and Philosophical Reflections on Modern Genetics,* ed. Robert F.

Weir, Susan C. Lawrence, and Evan Fales, 67–83. Iowa City: University of Iowa Press, 1994.

———. "'Our Load of Mutations' Revisited." *Journal of the History of Biology* 20, no. 3 (1987): 321–35.

———. *The Politics of Heredity: Essays on Eugenics, Biomedicine, and the Nature-Nurture Debate.* Albany: State University of New York Press, 1998.

Paul, Diane B., and Barbara Kimmelman. "Mendel in America: Theory and Practice, 1900–1919." In *The American Development of Biology,* ed. Ronald Rainger, Keith R. Benson, and Jane Maienschein, 281–310. New Brunswick, NJ: Rutgers University Press, 1988.

Pauling, Linus, H. A. Itano, S. J. Singer, and I. C. Wells. "Sickle Cell Anemia: A Molecular Disease." *Science* 110 (1949): 543–48. doi: 10.1126/science.110.2865.543.

Pauly, Philip. *Controlling Life: Jacques Loeb and the Engineering Ideal in Biology.* New York: Oxford University Press, 1987.

Pearl, Raymond. "Breeding Better Men." *The World's Work* 15 (1908): 9818–24.

Pearl, Raymond, Alan C. Sutton, William T. Howard, and Margaret G. Rioch. "Studies on Constitution: I. Methods." *Human Biology* 1 (1929): 10–56.

Pearson, Karl. *The Life, Letters, and Labours of Francis Galton.* 4 vols. Vol. 1, Cambridge: University Press, 1914.

Penrose, Lionel S. *A Clinical and Genetic Study of 1280 Cases of Mental Defect* [the Colchester Survey]. London: H. M. Stationery Office, Privy Council of Medical Research Council, 1938.

———. "Future Possibilities in Human Genetics." *American Naturalist* 76, no. 763 (1942): 165–70.

———. "Phenylketonuria: A Problem in Eugenics." *Lancet* 247, no. 6409 (1946): 949–53.

Pernick, Martin S. *The Black Stork: Eugenics and the Death of "Defective" Babies in American Medicine and Motion Pictures since 1915.* New York: Oxford University Press, 1996.

Pickens, Donald. *Eugenics and the Progressives.* Nashville, TN: Vanderbilt University Press, 1968.

Portelli, Alessandro. *The Death of Luigi Trastulli, and Other Stories: Form and Meaning in Oral History.* Suny Series in Oral and Public History. Albany: State University of New York Press, 1990.

Porter, Theodore M. *Karl Pearson: The Scientific Life in a Statistical Age.* Princeton, NJ: Princeton University Press, 2004.

Postle, Kathleen, and Arthur Postle. "Whose Little Girl Are You?" *McCall's,* December 1948, 22–23, 126, 136–38.

Proctor, Robert. *Racial Hygiene: Medicine under the Nazis.* Cambridge: Harvard University Press, 1988.

"A Proposed Standard System of Nomenclature of Human Mitotic Chromosomes (Denver, Colorado)." *Annals of Human Genetics* 24 (1960): 319–25.

Provine, W. B. "Geneticists and the Biology of Race Crossing." *Science* 182 (1973): 790–96. doi: 10.1126/science.182.4114.790.

Ptashne, M. "Isolation of the lambda Phage Repressor." *Proceedings of the National Academy of Sciences U S A* 57, no. 2 (1967): 306–13.

Rader, Karen A. *Making Mice*. Princeton: Princeton University Press, 2004.

Rainer, J. D. "Franz Kallmann's Views on Eugenics." *American Journal of Psychiatry* 146, no. 10 (1989): 1361–62.

Reed, Sheldon C. "A Short History of Genetic Counseling." *Social Biology* 21 (1975): 332–39.

———. "Reactivation of the Dight Institute, 1947–1949." *Dight Institute of the University of Minnesota* 6 (1949): 1–5.

Reilly, Philip. *The Surgical Solution: A History of Involuntary Sterilization in the United States*. Baltimore: Johns Hopkins University Press, 1991.

Reinhold, Robert. "Scientists Isolate a Gene; Step in Heredity Control." *New York Times*, November 23, 1969.

Resnik, D. B., and D. B. Vorhaus. "Genetic Modification and Genetic Determinism." *Philosophy, Ethics, and Humanities in Medicine* 1, no. 1 (2006): E9. doi: 10.1186/1747/5341/1/9.

Resta, R. G. "The Historical Perspective: Sheldon Reed and 50 Years of Genetic Counseling." *Journal of Genetic Counseling* 6, no. 4 (1997): 375–77.

Resta, R. G., and D. Paul. "Historical Aspects of Medical Genetics." *American Journal of Medical Genetics* 115, no. 2 (2002): 73–74.

Rettig, Richard A. *Cancer Crusade: The Story of the National Cancer Act of 1971*. 2nd ed. Washington, DC: Joseph Henry, 2005.

Ridley, Matt. Foreword to Witkowski, Inglis, and Davenport, *Davenport's Dream*, ix–xi.

Riis, P., and F. Fuchs. "Antenatal Determination of Fetal Sex: Prevention of Hereditary Diseases." *Lancet* 276, no. 7143 (1960): 180–82.

Riis, P., F. Fuchs, S. G. Johnsen, J. Mosbech, and C. E. Pilgaard. "Cytological Sex Determination in Disorders of Sexual Development." In Kemp, Hauge, and Harvald, *Proceedings of the First International Congress of Human Genetics*, 256–60.

Riordan, J. R., J. M. Rommens, B. Kerem, N. Alon, R. Rozmahel, Z. Grzelczak, J. Zielenski, et al. "Identification of the Cystic Fibrosis Gene: Cloning and Characterization of Complementary DNA." *Science* 245 (1989): 1066–73. doi: 10.1126/science.2475911.

Robbins, Emily F. "Proceedings of the First National Conference on Race Betterment, January 8, 9, 10, 11, 12, 1914. Battle Creek, Michigan (1914)."

Roblin, R. "The Boston XYY Case." *Hastings Center Report* 5, no. 4 (1975): 5–8.

Rodgers, Daniel T. "In Search of Progressivism." *Reviews in American History* 10, no. 4 (1982): 113–32.

Roll-Hansen, Nils. "Geneticists and the Eugenics Movement in Scandinavia." *British Journal for the History of Science* 22 (1989): 335–46.

Romanes, George. "Human Faculty and Its Development." *Nature* 28 (1883): 97–98. doi: 10.1038/028097a0.

Rommens, J. M., M. C. Iannuzzi, B. Kerem, M. L. Drumm, G. Melmer, M. Dean, R. Rozmahel, et al. "Identification of the Cystic Fibrosis Gene: Chromosome Walking and Jumping." *Science* 245 (1989): 1059–65, doi: 10.1126/science.2772657.

Roof, Judith. *The Poetics of DNA*. Minneapolis: University of Minnesota Press, 2007.

Rosen, Christine. *Preaching Eugenics: Religious Leaders and the American Eugenics Movement*. Oxford: Oxford University Press, 2004.

Rosen, George. "The Committee of One Hundred on National Health and the Campaign for a National Health Department, 1906–1912." *American Journal of Public Health* 62 (1972): 261–63.

Rosenberg, Charles E. "Piety and Social Action: Some Origins of the American Public Health Movement." In *No Other Gods: On Science and American Social Thought*, 109–22. Baltimore: Johns Hopkins University Press, 1976.

Rossiter, Margaret. "The Organization of the Agricultural Sciences." In *The Organization of Knowledge in Modern America, 1860–1920*, ed. Alexandra Oleson and John Voss, 211–48. Baltimore: Johns Hopkins University Press, 1979.

Ruddle, Frank H. Interview with Nathaniel Comfort, December 4, 2006, New Haven. Oral History of Human Genetics Project, UCLA and Johns Hopkins University, http: ohhgp.pendari.com.

Sanger, Margaret. *The Pivot of Civilization*. New York: Brentano's, 1922.

Sayre, Anne. *Rosalind Franklin and DNA*. New York: Norton, 1975.

Schieffelin, William Jay. "Work of the Committee of One Hundred on National Health." *Annals of the American Academy of Political and Social Science* 37, no. 2 (1911): 321–29.

Schmeck, Harold M., Jr. "Virus Is Injected into 2 Children in Effort to Alter Chemical Trait." *New York Times*, September 21, 1970.

Schmidt, B. C. "Five Years into the National Cancer Program." *Yale Journal of Biology and Medicine* 50, no. 3 (1977): 237–44.

Schneider, William H. "The Eugenics Movement in France." In *The Wellborn Science: Eugenics in Germany, France, Brazil, and Russia*, ed. Mark B. Adams, 69–109. New York: Oxford University Press, 1990.

Schoen, Johanna. *Choice and Coercion: Birth Control, Sterilization, and Abortion in Public Health and Welfare*. Gender and American Culture. Chapel Hill: University of North Carolina Press, 2005.

———. "Fighting for Child Health: Race, Birth Control, and the State in the Jim Crow South." *Social Politics* 4, no. 1 (1997): 90–113. doi: 10.1093/sp/4.1.90.

———. "Reassessing Eugenic Sterilization: The Case of North Carolina." In *A Century of Eugenics in America*, ed. Paul Lombardo, 141–60. Bloomington: Indiana University Press, 2011.

Schull, William J. "James Van Gundia Neel (1915–2000)." *Biographical Memoirs* 81 (2002): 1–21.

Schwartz, James. *In Pursuit of the Gene: From Darwin to DNA*. Cambridge: Harvard University Press, 2008.

Schwarz, Richard W. *John Harvey Kellogg, M.D.* Nashville, TN: Southern Publishing Association, 1970.

"Scientific Notes and News." *Science* 31 (1910): 781–83. doi: 10.1126/science.31.803.781.

Scriver, Charles R. Interview with Nathaniel Comfort, August 22–23, 2006, Montreal. Oral History of Human Genetics Project, UCLA and Johns Hopkins University, http://ohhgp.pendari.com.

Selden, Steven. *Inheriting Shame: The Story of Eugenics and Racism in America.* Advances in Contemporary Educational Thought Series. New York: Teachers College Press, 1999.

Skloot, Rebecca. *The Immortal Life of Henrietta Lacks.* New York: Crown, 2010.

Slater, L. B. "Malaria Chemotherapy and the 'Kaleidoscopic' Organisation of Biomedical Research during World War II." *Ambix* 51, no. 2 (2004): 107–34.

Smith, Michael. *Lionel Sharples Penrose: A Biography.* Colchester: Michael Smith, 1999.

Smith, Virginia. *Clean: A History of Personal Hygiene and Purity.* Oxford: Oxford University Press, 2007.

Smithies, Oliver. Interview with Nathaniel Comfort, Nov. 26–27, 2005, Chapel Hill, NC. Oral History of Human Genetics Project, UCLA and Johns Hopkins University, http://ohhgp.pendari.com.

———. "An Improved Procedure for Starch-Gel Electrophoresis—Further Variations in the Serum Proteins of Normal Individuals." *Biochemical Journal* 71 (1959): 585–87.

———. "Starch-Gel Electrophoresis." *Metabolism* 13 (1964): suppl. 974–84.

———. "Zone Electrophoresis in Starch Gels: Group Variations in the Serum Proteins of Normal Human Adults." *Biochemical Journal* 61, no. 4 (1955): 629–41.

Smyth, C. J., C. W. Cotterman, and R. H. Freyberg. "The Genetics of Gout and Hyperuricaemia." *Annals of the Rheumatic Diseases* 7, no. 4 (1948): 248.

———. "The Genetics of Gout and Hyperuricaemia: An Analysis of 19 Families." *Journal of Clinical Investigation* 27, no. 6 (1948): 749–59.

Snyder, Laurence H. *Blood Grouping in Relation to Clinical and Legal Medicine.* Baltimore: Williams and Wilkins, 1929.

———. "Genetics in Medicine." *Ohio State Medical Journal* 29 (1933): 705–8.

———. "Human Blood Groups and Their Bearing on Racial Relationships." *Proceedings of the National Academy of Sciences U S A* 11, no. 7 (1925): 406–7.

———. "Human Blood Groups: Their Inheritance and Racial Significance." *American Journal of Physical Anthropology* 9, no. 2 (1926): 233–63.

———. "Inherited Taste Deficiency." *Science* 74 (1931): 151–52. doi: 10.1126/science.74.1910.151.

———. "Linkage in Man." *Eugenical News* 18, no. 8 (1931): 117–19.

———. *Medical Genetics: A Series of Lectures Presented to the Medical Schools of Duke University, Wake Forest College, and the University of North Carolina.* Durham, NC: Duke University Press, 1941.

———. "The Principles of Gene Distribution in Human Populations." *Yale Journal of Biology and Medicine* 19, no. 5 (1947): 817–33.

———. *The Principles of Heredity.* Boston: Heath, 1935.

———. "The Study of Human Heredity." *Scientific Monthly* 51, no. 6 (1940): 536–41.

———. Telephone Interview with Daniel J. Kevles, April 29, 1983. Daniel Kevles Collection, Rockefeller Archive Center.

Snyder, Laurence H., R. C. Baxter, and A. W. Knisely. "Studies in Human Inheritance: XIX. The Linkage Relations of the Blood Groups, the Blood Types, and Taste Deficiency to P. T. C." *Journal of Heredity* 32 (1941): 22–25.

Snyder, Laurence H., and George M. Curtis. "An Inherited 'Hollow Chest': Koilosternia, a New Character Dependent upon a Dominant Autosomal Gene." *Journal of Heredity* 25 (1934): 445–47.

Solomon, E., and W. F. Bodmer. Introduction to *Cytogenetics and Cellular Genetics* 58, no. 1 (1991). doi: 10.1159/000133157.

Soyfer, V. N. "Tragic History of the VII International Congress of Genetics." *Genetics* 165, no. 1 (2003): 1–9.

Spiro, J. P. "Nordic vs. Anti-Nordic: The Galton Society and the American Anthropological Association." *Patterns of Prejudice* 36, no. 1 (2002): 35–48. doi: 10.1080/003132202128811358.

Stansfield, W. D. "The Bell Family Legacies." *Journal of Heredity* 96, no. 1 (2005): 1–3.

Steinberg, A. G. "Much Ado About Me." *American Journal of Medical Genetics* 59, no. 2 (1995): 250–62.

Steinfels, M. O., and C. Levine. "The XYY Controversy: Researching Violence and Genetics." *Hastings Center Report* 10, no. 4 (1980): suppl. 1–32.

Stent, Gunther S. *The Coming of the Golden Age: A View of the End of Progress.* Garden City, NY: Natural History Press (Doubleday), 1969.

"The Sterilization of Mental Defectives." *Journal of the American Medical Association* 95, no. 8 (1930): 605.

Stern, Alexandra. *Eugenic Nation: Faults and Frontiers of Better Breeding in Modern America.* American Crossroads. Berkeley: University of California Press, 2005.

———. *Telling Genes: The Story of Genetic Counseling in Modern America.* Baltimore: Johns Hopkins University Press, 2012.

Stolberg, Michael. "The Decline of Uroscopy in Early Modern Learned Medicine (1500–1650)." *Early Science and Medicine* 12, no. 3 (2007): 313–36.

Stolberg, S. G. "The Biotech Death of Jesse Gelsinger." *New York Times Magazine,* November 28, 1999, 136–40, 149–50.

Strandskov, Herluf H. "Discussion: Human Gene Symbols." *Science* 94 (1941): 366–67. doi: 10.1126/science.94.2442.366.

Strasser, Bruno J. "Linus Pauling's 'Molecular Diseases': Between History and Memory." *American Journal of Medical Genetics* 115, no. 2 (2002): 83–93.

Sturtevant, A[lfred]. H. "Genetic Factors Affecting the Strength of Linkage in Drosophila." *Proceedings of the National Academy of Sciences U S A* 3, no. 9 (1917): 555–58.

———. *A History of Genetics.* New York: Harper and Row, 1965.

Summers, William C. "How Bacteriophage Came to Be Used by the Phage Group." *Journal of the History of Biology* 26 (1993): 255–67.

Swazey, J. P., J. R. Sorenson, and C. B. Wong. "Risks and Benefits, Rights and Responsibilities: A History of the Recombinant DNA Research Controversy." *Southern California Law Review* 51, no. 6 (1978): 1019–78.

Sweeney, Gerald. *"Fighting for the Good Cause": Reflections on Francis Galton's Legacy to American Hereditarian Psychology.* Transactions of the American Philosophical Society 91, part 2. Phildelphia: American Philosophical Society, 2001.

Szreter, Simon. "The Importance of Social Intervention in Britain's Mortality Decline C. 1850–1914: A Reinterpretation of the Role of Public Health." *Social History of Medicine* 1 (1988): 1–38.

Szybalska, E. H., and W. Szybalski. "Genetics of Human Cell Line: IV. DNA-Mediated Heritable Transformation of a Biochemical Trait." *Proceedings of the National Academy of Sciences U S A* 48 (1962): 2026–34.

Szybalski, Waclaw, E. H. Szybalska, and G. Ragni. "Genetic Studies with Human Cell Lines." *National Cancer Institute Monographs* 7 (1962): 75–89.

Telfer, M. A., D. Baker, G. R. Clark, and C. E. Richardson. "Incidence of Gross Chromosomal Errors among Tall Criminal American Males." *Science* 159 (1968): 1249–50. doi: 10.1126/science.159.3820.1249.

Timofeeff-Ressovky, N. W., K. G. Zimmer, and Max Delbrück. "Über die Natur der Genmutation und der Genstruktur." *Nachrichten von der Gesellschaft der Wissenschaften zu Göttingen: Mathematische-Physikalische Klasse, Fachgruppe VI, Biologie* 1, no. 13 (1935): 189–245.

Tjio, Joe H., and Albert Levan. "The Chromosome Number of Man." *Hereditas* 42 (1956): 1–6.

Todes, Daniel Philip. *Pavlov's Physiology Factory: Experiment, Interpretation, Laboratory Enterprise.* Baltimore: Johns Hopkins University Press, 2001.

Tomes, Nancy. *The Gospel of Germs: Men, Women, and the Microbe in American Life.* Cambridge: Harvard University Press, 1998.

Tracy, Sarah W. "An Evolving Science of Man: The Transformation and Demise of American Constitutional Medicine, 1920–1950." In Lawrence and Weisz, *Greater than the Parts,* 161–88.

———. "George Draper and American Constitutional Medicine, 1916–1946: Reinventing the Sick Man." *Bulletin of the History of Medicine* 66 (1992): 53–89.

Tucker, William H. *The Funding of Scientific Racism: Wickliffe Draper and the Pioneer Fund.* Urbana: University of Illinois Press, 2002.

United States Office of Scientific Research and Development and Vannevar Bush. *Science, the Endless Frontier: A Report to the President.* Washington, DC: Government Printing Office, 1945.

Valle, David. "Victor Almon Mckusick, MD, in Memoriam." *American Journal of Human Genetics* 83, no. 3 (2008). doi: 10.1016/j.ajhg.2008.08.017.

Various. "Discussion of the Advisability of the Registration of Tuberculosis." *Transactions and Studies of the College of Physicians of Philadelphia* 16 (1894): 1–27.

Vogel, Friedrich. "Moderne Problem der Humangenetik." *Ergebnisse der inneren Medizin und Kinderheilkunde* 12 (1959): 52–125.

Wade, N. "Special Virus Cancer Program: Travails of a Biological Moonshot." *Science* 174 (1971): 1306–11. doi: 10.1126/science.174.4016.1306.

Wade, Nicholas. "Genome of DNA Discoverer Is Deciphered." *New York Times,* June 1, 2007.

Walker, J. C. "Putting Method First: Re-Appraising the Extreme Determinism and Hard Hereditarianism of Sir Francis Galton." *History of Science* 40, no. 127, part 1 (2002): 35–62.

Wallace, Bruce, and Theodosius Grigorievich Dobzhansky. *Radiation, Genes, and Man.* New York: Holt, 1959.

Waller, J. C. "Gentlemanly Men of Science: Sir Francis Galton and the Professionalization of the British Life-Sciences." *Journal of the History of Biology* 34, no. 1 (2001): 83–114.

Walzer, S[tanley]., G. Breau, and P. S. Gerald. "A Chromosome Survey of 2,400 Normal Newborn Infants." *Journal of Pediatrics* 74, no. 3 (1969): 438–48.

———. Interview with Ami Karlage, August 23, 2008, Newton, MA. Oral History of Human Genetics Project, UCLA and Johns Hopkins University, http://ohhgp.pendari.com.

Watkins, Elizabeth Siegel. *On the Pill: A Social History of Oral Contraceptives, 1950–1970.* Baltimore: Johns Hopkins University Press, 1998.

Watson, James D[ewey]. "In Further Defense of DNA." (1978).

———. "Moving toward the Clonal Man." *Atlantic Monthly,* May 1971, 53 http://www.theatlantic.com/magazine/archive/1971/05/moving-toward-the-clonal-man/5435/1/.

Watson, James D., and Gunther S. Stent. *The Double Helix: A Personal Account of the Discovery of the Structure of DNA.* New York: Scribner, 1998.

Watson, James D., and John Tooze. *The DNA Story: A Documentary History of Gene Cloning.* San Francisco: Freeman, 1981.

Watt, David C. "Lionel Penrose, F.R.S. (1898–1972) and Eugenics." *Notes and Records of the Royal Society of London* 52 (1998): 137–51, 339–54.

———. "L. S. Penrose (1898–1972): Psychiatrist and Professor of Human Genetics." *British Journal of Psychiatry* 173 (1998): 458–61.

Wegner, Phillip E. *Imaginary Communities: Utopia, the Nation, and the Spatial Histories of Modernity.* Berkeley: University of California Press, 2002.

Weisberger, A. S. "Induction of Altered Globin Synthesis in Human Immature Erythrocytes Incubated with Ribonucleoprotein." *Proceedings of the National Academy of Sciences U S A* 48 (1962): 68–80.

Weiss, Mary C., and Howard Green. "Human-Mouse Hybrid Cell Lines Containing Partial Complements of Human Chromosomes and Functioning Human Genes." *Proceedings of the National Academy of Sciences U S A* 58, no. 3 (1967): 1104–11.

Weiss, Rick, and Deborah Nelson. "Teen Dies Undergoing Experimental Gene Therapy." *Washington Post,* December 8, 1999.

Weiss, S[heila]. F[aith]. "After the Fall: Political Whitewashing, Professional Posturing, and Personal Refashioning in the Postwar Career of Otmar Freiherr von Verschuer." *Isis* 101 (2010): 722–58.

———. *Race Hygiene and National Efficiency: The Eugenics of Wilhelm Schallmayer.* Berkeley: University of California Press, 1987.

———. "The Race Hygiene Movement in Germany." In *The Wellborn Science: Eugenics in Germany, France, Brazil, and Russia,* ed. Mark B. Adams, 8–68. New York: Oxford University Press, 1990.

Weiss, Soma, and Laurence B. Ellis. "The Rational Treatment of Arterial Hypertension." *Journal of the American Medical Association* 95, no. 12 (1930): 846–52.

Weldon, W. F. R. "The Study of Animal Variation." *Nature* 50 (1894): 25–26. doi: 10.1038/050025a0.

Weller, Thomas H., and Frederick C. Robbins. "John Franklin Enders (1897–1985)." *Biographical Memoirs* 60 (1991): 47–65.

Wexler, Alice. *Mapping Fate: A Memoir of Family, Risk, and Genetic Research.* New York: Times Books, Random House, 1995.

Whitney, David Day. *Family Treasures: A Study of the Inheritance of Normal Characteristics in Man.* Humanizing Science Series. Lancaster, PA: Jacques Cattell, 1942.

Whorton, James C. *Nature Cures: The History of Alternative Medicine in America.* Oxford: Oxford University Press, 2002.

Wiebe, Robert H. *The Search for Order, 1877–1920.* Making of America. New York: Hill and Wang, 1967.

Wiener, A. S. "Genetics and Nomenclature of the Rh-Hr Blood Types." *Antonie von Leeuwenhoek* 15, no. 1 (1949): 17–28.

———. "Genetic Theory of the Rh Blood Groups." *Proceedings of the Society for Experimental Biology and Medicine* 54 (1943): 316–19.

———. "Nomenclature of the Rh Blood Types." *Science* 99 (1944): 532–33. doi: 10.1126/science.99.2583.532.

Williams, Roger John. *Biochemical Individuality.* New York: Wiley, 1956.

———. *Why Human Genetics? Journal of Heredity* 51: 91–98, 1960.

Wilson, Philip K. "Confronting 'Hereditary' Disease: Eugenic Attempts to Eliminate Tuberculosis in Progressive Era America." *Journal of Medical Humanities* 27, no. 1 (2006): 19–37.

———. "Harry Laughlin's Eugenic Crusade to Control the 'Socially Inadequate' in Progressive Era America." *Patterns of Prejudice* 36, no. 1 (2002): 49–67. doi: 10.1080/003132202128811367.

———. "Pedigree Charts as Tools to Visualize Inherited Disease in Progressive Era America." In *A Cultural History of Heredity: Heredity in the Century of the Gene,* ed. S. Müller-Wille and H.-J. Rheinberger, 163–90. Berlin: Max-Plank-Institut für Wissenschaftsgeschichte, 2008.

Witkowski, J. A. "Charles Benedict Davenport, 1866–1944." In Witkowski, Inglis, and Davenport, *Davenport's Dream*, 36–58.

Witkowski, J. A., J. R. Inglis, and Charles Benedict Davenport. *Davenport's Dream: 21st Century Reflections on Heredity and Eugenics*. Cold Spring Harbor, NY: Cold Spring Harbor Laboratory Press, 2008.

Wolfe, Audra. "The Organization Man and the Archive: A Look at the Bentley Glass Papers." *Journal of the History of Biology* 44, no. 1 (2011): 147–51. doi: 10.1007/s10739-011-9276-6.

Wolkow, Michael, and E. Baumann. "Ueber das Wesen der Alkaptonurie." *Zeitschrift für physiologische Chemie* 15 (1891): 228–85.

Wright, Susan. "Recombinant DNA Technology and Its Social Transformation, 1972–1982." *Osiris*, 2 (1986): 303–60.

Yater, Wallace M., and Mario Mollari. "The Pathology of Sickle-Cell Anemia." *Journal of the American Medical Association* 96, no. 20 (1931): 1671–75.

Yi, D. "Cancer, Viruses, and Mass Migration: Paul Berg's Venture into Eukaryotic Biology and the Advent of Recombinant DNA Research and Technology, 1967–1980." *Journal of the History of Biology* 41, no. 4 (2008): 589–636.

Zallen, D. T., and R. M. Burian. "On the Beginnings of Somatic Cell Hybridization: Boris Ephrussi and Chromosome Transplantation." *Genetics* 132, no. 1 (1992): 1–8.

Zinder, Norton, and Joshua Lederberg. "Genetic Exchange in Salmonella." *Journal of Bacteriology* 64 (1952): 679–99.

INDEX

Note: Page numbers in **boldface** refer to illustrations.